A Guide to
EFFECTIVE INDUSTRIAL SAFETY

Gulf Publishing Company, Book Division, Houston, Texas

A GUIDE TO EFFECTIVE INDUSTRIAL SAFETY

JACK W. BOLEY

*To Joyce, whose encouragement
made this book possible*

A Guide to
EFFECTIVE INDUSTRIAL SAFETY

Library of Congress
Catalog Card Number:
76-23915

ISBN: 0-87201-798-2

Contents

Foreword

"He couldn't see the forest for the trees." This old cliché fits well the current health and safety scene in America and in most industrialized nations. Legislation has brought to bear tremendous pressures on safety professionals and managers. It has immersed them in a growing mountain of detail and paper work . . . OSHA forms and records . . . ever-more-stringent codes and standards . . . an endless flow of toxic-substance data sheets and the like. A morass of detail rules the safety administrator's life.

How easy it is, in the midst of this complexity, to concentrate our safety awareness effort on a "hot" topic that happens to be the subject of some new bureaucratic demand. The temptation is always there to take on a "brush fire" mentality, and thus risk missing the "big picture." The trees obscure the forest. One could even construe such narrowness as being *unsafe*.

For all these reasons and more this book, which might otherwise have been criticized for being too "broad and general," fills a great need. Simply and succinctly it places into perspective most of the safety/health areas of which safety professionals should be well informed, and upon which they should be prepared to act. For the seeker of details, the author has wisely included an appendix which lists organizations and publications that offer boundless detail on a great variety of health/safety topics. This well-balanced, basic book should be a welcome addition to the health/safety literature. It is fundamental enough for the student and comprehensive enough to provide the "old pro" with a neat wrap-up of what the effective practice of industrial safety is all about.

Charles Vervalin
Fire/Safety Editor
Hydrocarbon Processing magazine

1
Management
Accountability

For any organization to function successfully someone has to be held accountable. The safety effort in an organization is no different than production, quality control, or sales. Each of these functions must be led by a responsible individual who can ultimately be held accountable for its success or failure.

Much has been written in the past few years about management participation in the company safety effort. No doubt this participation is essential if a company is to achieve and maintain a high level of safety performance. The point of contention is that many qualified individuals, as well as some recognized safety consultant groups, insist that top management must develop and sign a written safety policy, if top safety performance is to be accomplished. In many cases this does mean a total commitment to the safety effort by top management. However, in many cases the safety policy is written by someone in the safety department and the top executive is convinced that it should be signed because, if nothing else, it looks good to anyone that may question the safety support of management. However, if management, in signing the safety policy, does not fully intend to support the policy with the needed finances and actively participate in the safety effort, they should not issue the policy. A policy that has been signed and issued, but not supported has a degrading effect on the entire safety effort as soon as the employees discover the policy is only so many words on paper.

If safety is to take its proper place along the side of production and quality control, there cannot be a separate safety policy, just as there cannot be separate policies on production and quality control. It is much better to develop an overall management philosophy where safety is a normal part of all operations. Each level of management must ensure that their immediate subordinates understand their safety responsibility, are trained to carry out this responsibility, and are held accountable for the results.

Safety and the First-Line Supervisor

Safety, as well as production and quality control, ultimately are the responsibility of the first-line supervisor. This individual usually has more jobs than he can properly handle. It is of upmost importance that these individuals possess the proper training to handle these responsibilities.

It is felt by many that the simpler the approach to safety, the better the safety performance. If a supervisor is saddled with a continuous process of developing safe work procedures, awards programs, and other similar activities, he may at the first opportunity neglect some or all of them. If this situation occurs, the safety department should give additional help and guidance to the supervisor. Certainly the supervisor must participate in maintaining safe work procedures and a high standard of safety performance is expected. In order to achieve this objective, the first-line supervisor must have his responsibility and accountability fully defined. This involves knowing the answers to the following questions:

1. What is his safety responsibility and authority? How is it integrated into his total responsibility?
2. What is to be done about maintenance and repairs?

3. What is his responsibility for determining qualifications of workers? What disciplinary action is permitted? Under what circumstances?
4. What are the safe work methods for each job under his responsibility?
5. What commitments may he make to correct unsafe conditions and what maximum expenditure may he make without additional authority?

One of the most difficult, yet most important, safety activities for the supervisor to develop and maintain is a good safety attitude in the minds of all employees. When a good safety attitude is maintained, employees will perform their work safely without close supervision. Good safety attitudes normally will ensure efficient safe operations, yet poor safety attitudes will cause accidents in operations that under normal conditions would be accident free.

How a Safety Program Fails

A supervisor that develops a program of safe attitude building, morale building, and safety training has a program that is far superior to one that is only based on mottoes, contests, or awards programs. It is not to say that these devices when used as part of a well-balanced safety program cannot be effective. However, in most cases where a program is based on these devices alone, it is ultimately doomed to failure.

Usually a safety contest will get off to a roaring start with a lot of publicity only to die a slow agonizing death. This happens when the employees realize that they are participating in a carnival type side show and no real effort is being made to improve working conditions or to train employees in accident prevention. This realization can often cause the accident picture to be worse than it was before the contest was started. This can happen in several ways, for instance when a contest is based on lost time injuries, employees may try to hide an injury to preserve the safety record or win the contest. The problem begins when an injury is not treated and becomes infected, or the small injury is further injured while the employee is trying to perform his work in the injured condition. Also, when the employees realize that no real effort to improve the safety program is being made, a "backlash" might occur which can shift the employees' safety attitude from casual indifference to premeditated negligence. This results in a substantial increase in accidents.

When a contest or awards program is part of a well-balanced safety program, it can be effective. When this approach is used, it must be based on long term meaningful accomplishments. Also, the goals that are set must be achievable. Care must be taken to divide work groups into workable units. Only competition within each unit shall be allowed, otherwise undesirable events, such as the safety contest previously described, can occur. When the preset goal is reached, management should immediately provide the award, whatever it may be.

When an awards program or contest is initiated care should be taken to ensure a proper award. The safety dinner is one of the most popular awards in many industries. It can be enjoyable if handled correctly. However, there is a tendency to have a safety dinner which includes alcoholic beverages. Care must be taken to limit the amount of alcohol employees consume, as it is very poor management judgment to bring employees together for a safety awards dinner then send them driving home half drunk. If an unlimited amount of alcohol is to be available, then some preset arrangements must be made where the employees will not be driving automobiles at the conclusion of the event.

Safety Committees

Another management tool that can be used to improve safety performance is a safety committee. However, it should be used with discretion. There are several types of safety committees, but the following are the most common:

1. Management safety committee—usually made up of six or eight top managers in the plant.

2. Management union safety committee—made up of six or eight employees, equally divided between management and labor.
3. Shop safety committee—usually consists of the shop supervisor and three or four employees.

Whatever the makeup of the safety committee, they are usually effective only in the areas of improving working conditions and eliminating physical hazards, which many authorities feel only cover about 20% of the real safety problems.

If these committees are not closely monitored, they may cause the available safety dollar to be spent unwisely. Sometimes emotions guide these committees instead of reason.

It is felt by many knowledgeable safety professionals that the safety committee is not needed and is actually undesirable. The alternate to the safety committee is well-trained management from top to bottom that assumes and discharges their safety responsibilities. These safety professionals feel that management should not delegate their safety responsibilities to committees anymore than they should delegate quality, production, or financial decisions.

Safety Performance Evaluation

The "Lost-Time-Injury-Cover-Up" Syndrome

Management in many organizations tries to evaluate safety performance on the basis of lost time injuries alone. This is a mistake because it causes undue pressure on each location to reduce lost time injuries, yet reportable injuries or doctor cases are not considered. This type of pressure invariably causes many managers to cover up lost time injuries. When this happens two things will occur, first the employees will lose respect for the safety effort when they see that local management will go to any length to preserve their safety record. Secondly, it gives top management the false impression that the safety performance is in much better condition than it really is. This puts local management in a very precarious position when they are trying to increase their safety budget to help their safety performance. Many case histories of safety engineers having difficulty in obtaining top management support can be traced directly back to a "lost-time-injury-cover-up" syndrome. If top management has been convinced by accidents being covered up that all is well, they honestly think they are giving all the monetary support to safety that is needed.

Another direct cause of the lost-time-injury-cover-up syndrome is a safety contest. It is a well known fact that the governing standard is loosely interpreted by many individuals simply to win a safety contest or achieve "x" number of man-hours without a lost time injury. Anyone or any organization that sponsors safety contests may unwillingly aid or cause a lost time injury cover-up. Again, a direct result of the cover-up is the workingman's contempt for something so false. He feels the company really isn't concerned that he may be injured on the job.

Many actual case histories have been uncovered where a company or plant has achieved a significant number of man-hours without a lost time injury, yet it is well known local fact that a substantial cadre of walking wounded are available for various nonproductive work tasks.

The cover-up can take many forms. Listed below are some that have been encountered the past few years:

1. A maintenance man breaks his arm. The next day he is put on a guard post; no lost time.
2. A woman slips on ice outside of plant door. The safety department takes the attitude that the accident could not have been prevented; the accident was just not counted.
3. A man strains his back on a hard to open valve. The company doctor says the man's physical condition caused the accident; it is not counted.
4. A man slips in shower at the end of shift and breaks his arm. The attitude is taken that productive work was not being performed; the accident is not counted.

5. An office worker has an allergic reaction from paint fumes while office is being repainted. She takes the rest of the day off, comes in the next day, which is Friday, only to empty a waste basket, and goes home; the problem is over by Monday. Occupational illness is not counted.

This type of cover-up history is endless, and the examples are from various industries, from many different parts of the country. This seems to be a wide spread blight on the safety movement.

Another problem associated with judging a safety performance solely on lost time injuries is in the way of comparison. Each plant or location is probably being judged against national averages or another part of the company. The problem here is that it is not realized that many of these statistics are bogus because of the lost time injury cover-up.

Many leaders in the safety field think that management should use lost time injuries as no more than 20% of the safety performance evaluation. There are many factors that should also be used in the safety performance evaluation. Listed below is a breakdown of how some organizations evaluate their safety performance:

1. Lost time injuries, 20%.
2. Safety meetings, 20%.
3. Accident investigations, 20%.
4. Supervisor safety training, 20%.
5. Facility inspection, 10%.
6. Off-the-job safety, 5%.
7. Awards program, 5%.

Safety Meetings

Using safety meetings as part of the safety performance evaluation is very important. In many safety meetings the only thing accomplished is the gathering of employees who are not too busy to come to the meeting. The supervisor may remember no more than five minutes before the safety meeting that he must conduct some type of safety meeting. No planning was done, no thoughts as to what needed to be discussed and only a half-hearted effort is made to get all employees to the meeting. This type of meeting is worthless, in fact, it should not be held under these circumstances. The meeting should be well planned, and topics should be selected and discussed which will be meaningful to the group. All attendees should be encouraged to actively participate in the meeting.

If meetings are scheduled in advance, and they should be, only extreme emergencies should prevent the meeting from being held. All employees should attend, and a record should be made of attendance and subject discussed. All of the above criteria should be used in evaluating the safety meeting. It is well to develop an evaluation system, where each point is assigned a value. A system such as this makes it much easier to evaluate the safety meeting. Figure 1-1 is a typical safety meeting performance evaluation sheet.

Accident Investigations

Accident investigations also must be a part of the safety performance evaluation. Much information can be gained to help prevent recurrences of similar accidents when a complete accident investigation is conducted. These are several points that should be determined in this evaluation:

1. Are all accidents investigated?
2. Are incidents (non-injury accidents) investigated?
3. Does the supervisor perform the accident investigation?
4. Does a formal supervisory accident report exist?
5. Is the report complete?
6. Were the necessary corrective actions taken?
7. If corrective actions were not taken, why?

Safety Meeting Performance Evaluation Sheet

Supervisor _____ Date _____
Subject of Meeting _____

		Points
1. No. of employees supervised _____		5-10-15-20
2. No. of employees in meeting _____		5-10-15-20
3. Was meeting scheduled in advance? _____		5-10-15-20
4. Was meeting held on day and time of schedule? _____		5-10-15-20
5. Were subjects meaningful to the group? _____		5-10-15-20
6. Were suggestions or recommendations received from the group, if so was a follow-up made? _____		5-10-15-20
	TOTAL	_____

Each of these questions should be answered, and points should be given to indicate the quality of each answer. Any point total of less than 80 should not be acceptable and the supervisor should be encouraged to improve the meetings.

Figure 1-1. *The effectiveness of a safety meeting can be determined by a performance evaluation sheet.*

8. Was the completed report seen by the next level of supervision?

All of these questions should be answered in the affirmative. If they are not, action should be taken by the local management to ensure that future accidents will be investigated in this manner.

Supervisory Safety Training

Supervisory safety training is so essential to good safety performance that it must be considered in the safety performance evaluation. One should look for the content of the training, how frequently it is given, how sincere management is about backing up the lessons taught in the training, and the supervisor's attitude about the training.

Facility Inspections

Facility inspections in many plants receive an excessive amount of manpower. They are needed, but much of the facility inspection effort could be utilized elsewhere, if a system of inspection and follow-up is rigidly enforced. Important points to look for when evaluating this area are, is the inspection written up formally?, how thorough does the report appear?, and how well is the follow-up being enforced?

Off-the-job Safety

Every safety program to be complete should include an element of the off-the-job safety. Certainly discretion has to be used when attacking the off-the-job safety problem. But it also must be recognized that in many cases the severest injuries, the most lost time, and the greatest expense experienced by a plant are caused by off-the-job injuries. So with this in mind it is felt that part of the evaluation should include off-the-job injuries.

Awards Program

The last point to cover is the awards program. It is not rated very high in an overall evaluation, but if used wisely and well planned it can help the overall safety performance of the location or plant.

If in fact the management sincerely wants to improve safety performance, the place to start is an honest accurate set of accident records. By admitting that a problem does exist, instead of trying to paint a rosy picture; by not covering up accidents; and by using a safety performance evaluation system instead of basing performance solely on lost time injuries, steps can be taken to start improving the overall safety performance.

2
The Loss Control Department

Any organization regardless of its size needs to implement a plan and make someone responsible for controlling human and material asset losses. A small ten-man operation certainly doesn't call for the manpower or sophistication that a one-thousand man operation requires. There are many factors that determine the type of loss control program a company should institute. They include such items as (1) size of operation; (2) location of operation, such as within a large industrial city or suburban location; (3) whether the operation is a sales office or a heavy manufacturing plant; (4) whether the operation is heavily dependent upon manpower or automatic equipment and machinery; and (5) the degree of hazard associated with the particular type of industry.

Safety Manager

Having the right person responsible for the safety program is the most important single factor in the entire program. Just as in any other function in the organization, or even more so, bad or weak leadership produces poor performance. Many top executives have a limited understanding of functions of the safety department, consequently very little time is spent in choosing the person to head the department. In more cases than most true safety professionals like to admit, this person is less than competent. Many organizations determine who is to head the group by one of the following methods: (1) the "good old boy" whom everyone likes yet can't seem to make it in any other department; (2) some poor fellow who was permanently injured in a company accident; or (3) some hack company politician who is all talk and no action.

None of these methods qualify the individual to head a safety department. In fact, the fellow who was in the accident probably is the only one who stands much of a chance of developing into the type of man needed to be in charge of the department. One only has to attend a large safety conference to see shining examples of the three types listed. Certainly it is not suggested that the majority of safety professionals fit into these types, but the safety effort certainly has more than its share.

The apparent question now is, what type of person should be placed in charge of the safety department? Of course, there are no perfect descriptions or qualifications. However, one should look for the same high quality individual as they would look for to head other profit generating departments in the company. It must be realized that the safety effort can be a profit generating department. Its profits may not be as easily identified as other departments because most organizations do not have a history or method of charging accident costs.

The person appointed to head the safety effort should possess most of the following qualifications: (1) the individual should be a person who gets along well with and enjoys being around his fellow man; (2) the individual should be as well-educated as the average manager in the organization; (3) he must possess as much leadership as the average manager in the organization; (4) he must be an innovator; (5) he must be ambitious and energetic; (6) he should have previous safety experience; (7) he must care about safety above all else.

Certainly each organization should modify the above qualifications to fit their particular need. However, the closer an organization stays to these qualifications the more successful the safety effort is likely to be.

One major weakness exists in many safety departments because top management makes a mistake in appointing an inefficient or unqualified individual to head the department, and then does not recognize the problem or is too slow in taking the necessary actions to correct the situation. There are many symptoms that top management should look for when determining if this condition does exist: (1) The department develops very few, if any, constructive programs. (2) The department head is constantly going to conventions, seminars, or other time-consuming functions. Certainly everyone needs to participate in some of these functions. It is the excessiveness that is pointed out here. (3) When you ask a question about a safety problem or anything that requires current knowledge, you end up ten minutes later, after a confusing dissertation, not knowing any more about the subject than you did when you asked the question. (4) Company equipment and man-hour losses stay at the previous level or continue to rise. (5) Policies and procedures are so vague, so poorly, or loosely written that no one really understands what is expected.

If these or other symptoms do indicate that the department is not functioning as it should, the department head should be looked at closely. If he is the problem, immediate steps should be taken to correct the situation.

Safety Functions

Up to this point the discussion is applicable to most organizations. However, four definite types of safety functions need lengthy discussions. They are corporate or divisional safety groups, one location or plant function, part-time functions, and loss control consultants.

Corporate or Divisional Safety Groups

The structure of the corporate or divisional safety group may vary of course, but to be effective certain structural elements

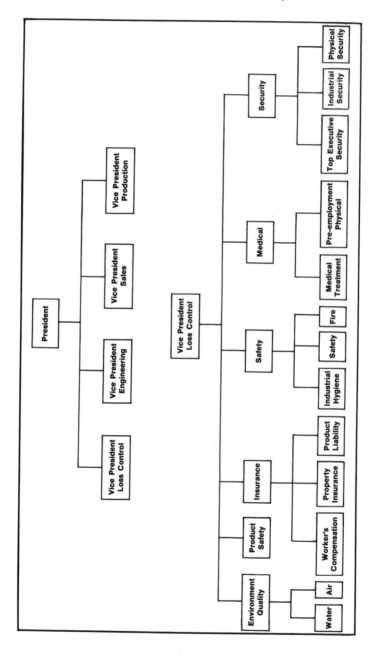

Figure 2-1. This chart indicates the position of safety in an ideal corporate structure.

should be utilized. (See Figure 2-1.) The organizational structure shown in Figure 2-1 is considered ideal, but is rarely used in actual practice. However, many organizations use a modified variation. The significant feature of the structure is that the head of the loss control group is a vice president. Putting loss control on a vice-presidential level is a fairly recent development. However, with the increase in consumerism, environmental impacts, federal regulations in safety, and the world-wide trend of corporate blackmail from terrorist groups, it is both desirable and necessary to have these functions reporting at a decision making level. With this type of structure, safety receives equal status with production, quality control, and sales.

The corporate or divisional safety department has a major responsibility to direct and guide the safety effort in all operational areas. The department should develop policies that reflect the company's desire in safety. These policies should be developed in conjunction with the safety specialist in each operational area. Once the policies have been adopted, the corporate staff should give the necessary support, such as training and orientation, to help the operational safety specialists implement the policies in their particular area. Some of these policies will cause local procedures to be developed. Again the corporate staff should, if necessary, lend a hand in their development.

Safety Audit

Another major function of the corporate staff is to conduct a safety audit of each operation at least semiannually. This audit should be a formal written report that attempts to evaluate the operation and point out any weaknesses in it. An example of such a safety audit is shown in Figure 2-2. Suggestions should be made as to how the weaknesses can be corrected. The written report should be given to the individual in charge of the operational area. It should be pointed out that deficiencies in the operation do not necessarily reflect on the safety specialists. It may, in fact, reflect directly on the manager of the operational area.

(Text continued on page 18)

Safety Audit

Plant or Location _____ Date _____
Manager _____ Safety Specialist _____

1. Date of last audit _____
2. Number of discrepancies _____.
3. Number of uncorrected discrepancies _____.
4. Why have discrepancies not been corrected? _____
5. Number of first aid injuries since last audit _____.
6. Number of OSHA reportable injuries since last audit _____
7. Number of lost time injuries since last audit _____.
8. Does a management investigation team investigate serious injuries? _____.
9. Number of OSHA reportables investigated by supervisors _____.
10. Number of follow up investigations by safety _____.
11. Quality of supervisor investigations—poor, fair, good, excellent.
12. Comment on quality _____.
13. Are pre-employment physicals given? _____.
14. Does an audiometric testing program exist? _____.
15. If yes, how many since last audit? _____.
16. Are specialized physicals given? _____.
17. If so, in (16) what kinds? _____.
18. What is frequency of above physicals? _____.
19. How many specialized physicals since last audit? _____
20. Does location have a nurse? _____.
21. Condition of first aid facility. Poor, fair, good, excellent.
22. Are lockout-tag out procedures in compliance? _____.
23. Is hot work permit procedure in use? _____.
24. Is entry permit procedure in use? _____.
25. Does spot check of work areas reveal compliance with procedures in (22), (23) and (24)? _____.
26. Condition of plant housekeeping _____.
27. Are safety work orders worked promptly? _____.
28. What dates are on the three oldest work orders?

_____ _____ _____.

29. Reason for any safety work order being over 60 days old.

_____

30. Does maintenance and new construction meet OSHA standards? _____.
31. Is personal protective equipment available for all needed jobs? _____ If no, why? _____
32. Is personal protective equipment used on all necessary jobs? _____ If no, why? _____
33. Are employees properly trained to use the required personal protective equipment? _____.
34. Are bulletin boards available? _____. Used? _____ Current? _____.

Figure 2-2. At least semiannually, a comprehensive safety audit such as this should be conducted of each operational area.

35. Do supervisors hold regularly scheduled safety meetings? _____.
36. What is quality of meetings (look at records and take random sample of employees)? _____.
37. How often is supervisory training held? _____.
38. Quality of supervisory training _____.
39. Does management conduct periodic facility inspections? _____. If yes, how often? _____.
40. Is the necessary follow-up provided? _____.
41. Is safety performance a part of the supervisory appraisal system? _____.
42. Is an employee's safety record considered before a promotion is given? _____.
43. Does safety appear on most management meetings agenda? _____.
44. Does a new employee safety training, and indoctrination program exist? _____. If yes, is it effective? _____.
45. Does a fire brigade exist? _____.
46. Is a fire detection system available on off shifts? _____.
47. Is detection system automatic or manual? _____.
48. Are fire extinguishing systems automatic? _____.
49. What type are the automatic systems? _____.
50. How many automatic systems? _____.
51. Did functional check of automatic systems operate as designed? _____.
52. Do fire pumps start automatically? _____.
53. How often are fire pumps started? _____.
54. How long do pumps run on each start? _____.
55. Condition of portable fire extinguishers? _____.
56. Are extinguisher records current? _____.
57. Are policies and procedures written in accordance with corporate desires? _____.
58. Any new local policies or procedures? _____.
59. If so, how many? _____. Subject Matter. _____.
60. Does plant safety staff attend seminars, schools and safety conferences? _____.
61. If so, what kinds? _____.
62. How many seminars, schools, and safety conferences since last audit? _____.
63. Does safety staff review maintenance and construction plans? _____.
64. Does the plant have an industrial hygiene program? _____.
65. What toxic substances are in the plant that require OSHA recordkeeping and maintaining? _____.
66. Are above records current? _____.
67. If above records are not current, why? _____.
68. Any current toxic substance problem areas? _____.
69. If a problem exists in (68), what? _____.
70. Walk around inspection. (Make list and write recommendations).
71. Overall plant evaluation. (Write summary).

Figure 2-2. Continued.

It should be a written corporate policy that the audit will occur and that the report will also be given to the loss control manager and to the operational manager's immediate supervisor. When the report reaches the loss control manager, it should be studied and then immediate decisions should be made as to what action or pressures are needed, if any, to bring the situation into the desired safety posture.

Corporate Fire Specialist

The corporate staff should review all plans of major new construction for fire prevention and extinguishing plans. More emphasis should be placed on prevention than extinguishment. It seems to be a current weakness throughout industry to rely on plans to control fires instead of fire prevention. It is not uncommon to see elaborate fire systems installed, yet not one flammable gas detector is in sight and there are no specifications of noncombustible materials in critical areas. These are the types of problems that should be eliminated by corporate reviews, if they exist. Policies and safety engineering specifications should exist to prevent this type of inferior construction. If the corporate staff does not have a qualified fire protection specialist, the services of a qualified consultant should be utilized in the review.

Industrial Hygienist

Each operation should have the services of a qualified industrial hygienist. Due to the short supply and high salaries it may be advisable for the corporate staff to include an industrial hygienist instead of placing one in each operational area. This, of course, depends on the work load and if the current governmental trend continues, a hygienist on the corporate staff may be overloaded to the point that hygienists will be required in certain operational areas. Although industrial hygiene may seem simple to the layman it is very tedious and exacting work that may require many man-hours to complete.

Seminars

The corporate staff should develop seminars for all the local safety specialists to attend at least annually. It is a good idea to rotate the location of these meetings from the corporate safety office to each operational area. One day should be set aside for plant or field tours to let the operational safety specialist experience each other's problems. The corporate staff may also recommend that various safety specialists attend a specialized school for further professional development.

Accident Costing

Another function of the corporate staff is to develop a program of accident cost. It is a well known fact that the medical payments and workman's compensation is only a small percentage of the actual cost of an accident. In fact, in many instances an accident does not result in a personal injury. Very little attention is paid to this type of accident in most companies. Hardly a day goes by that one does not read in the press of some industrial complex suffering major damage, yet no one sustains any injury. In far too many of these accidents few or inadequate investigations of the causes and costs are conducted. Too many local operating areas within a corporation try to minimize or play down the actual damage. When developing accident cost one must consider the following items:

1. Medical payments
2. Compensation payments
3. Overtime payments for replacement workers
4. Production delays
5. Product or material damage
6. Training of replacements
7. Accident investigation cost (which may include cost of bringing in a specialist in some cases)
8. Building or complex damages
9. Equipment damages
10. Business interruptions

When all of these items are added up in each accident, one can readily see that in a large percentage of the cases the medical and compensation payments are only a minor part of the overall costs.

If a company is to maintain a functioning loss control program, it is a necessity to include accident costing. Once the costs are presented to management, more than likely a much higher level of interest in loss control will be generated.

Accident Investigation

A companion to accident costing is complete and accurate investigations of all major injury and property damage accidents. Normally it is best to have a team headed by a corporate safety staff member investigate the accident. Usually the team should consist of three to five members, from various operating areas. The team should assemble and be at the accident scene as soon as possible. If possible, the accident scene should be placed under guard and left undisturbed.

Upon arrival at the accident scene, the team should obtain a complete set of pictures of the accident. It is almost impossible to take too many pictures. These pictures may be invaluable later in the investigation.

Statistical Data

The corporate staff should also assemble monthly an accident statistical report. This report, which should include OSHA reportable accidents, lost time injuries, deaths, and monthly accident cost, will help spot developing trends. The report should be sent to the president and all vice presidents so that they may keep abreast of the safety status in each area. It alone should not be used to judge the total safety performance of an operating area, although it probably is a good barometer.

A plant or operational area may require one or many safety specialists, depending upon the type of operation being conducted. The safety specialist in an operational area may take on many of the same duties of the loss control department, however, on a much smaller scale. Most plants have the plant nurse reporting to the safety specialist. The nurse in turn will normally file workers' compensation claims. In some plants the plant guards and environmental monitoring fall under the safety specialist's supervision. The major problem here is that one person is asked to perform many duties. There is a good chance that the safety specialist is not qualified for many of these duties. When this happens the whole safety and loss control effort will suffer, and management and employees will soon recognize the incompetence in the safety department. At this point respect and confidence are lost in the entire group, yet all or many in the group may be very competent in their respective field. There seems to be two ways out of this type dilemma. First, place the functions that the group is not qualified to perform in some other department or hire additional qualified employees. Second, if the staff already has enough employees, then each member should be given the necessary training to ensure that he or she can perform all the different functions.

The safety department (see Figure 2-3) should report to the plant or operational area manager. If the department reports at a lower level there is a feeling that the company is not placing as much emphasis on safety as it does on production, quality, and sales. The safety department should also report on a dotted line or functional basis to corporate safety for technical guidance. These reporting relationships help develop a continuing dialogue between the two areas. If both departments function as they should, much "reinventing the wheel" can be avoided.

Safety Department Operations

There seems to be three distinct methods of operating a safety department in a plant. (1) The safety department is strictly a staff function where members develop policies and procedure, make periodic inspections, and make recommendations in staff meetings. The line supervisor is totally responsible for the safety in his operation. (2) The safety department is a combination of staff and line functions. The department does all of the

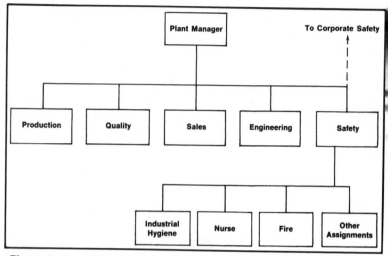

Figure 2-3. *As this chart shows, the safety department should share the same status as other departments, and should report to corporate safety on a functional basis.*

above as well as help out in line functions. They may help out in a particularly hazardous job or on major repair or construction they may issue permits. Sometimes they will hold safety meetings or perform other duties if time dictates. (3) The safety department acts strictly in a line function. They hold all safety meetings, investigate all accidents, issue all work permits, and perform any other function considered by anyone as a "safety job".

The business world has proponents of all three methods, however, many safety professionals feel the staff and line combination is the most workable. It lets the safety group become more involved in daily operations, yet it does not completely tie the department down such as the straight line function. In the straight staff function the main complaint is that the safety staff is cold and aloof, they don't really know what is going on in the plant. Certainly, praise and criticism of all three methods are justified because one method may lend itself to a particular industry better than another.

Staffing requirements in any plant depend on size, nature of business, hazard level of the business, location of plant in relation to municipal fire departments, and operating methods used by the safety department, such as staff, staff and line, or line. The staff method most likely will take the smallest safety staff and obviously the line should require the largest staff. Too many times an organization expects line performance of the safety group, yet provides only enough safety specialists to perform a staff function.

The plant safety staff should develop policies and procedures consistent with corporate objectives. This work should be done jointly with supervisors of the affected areas. When the supervisors are involved in the development of the policies and procedures that affect them, there is a much greater chance that compliance will be achieved.

The supervisor should receive safety training from the plant safety staff, and the corporate staff should help out if needed. Once the supervisor is completely trained and is aware of his safety responsibilities, he should be held accountable. Too many organizations do not train or only partially train their supervisors, yet hold them accountable for safety. This causes low morale among the supervisors as well as poor safety performance. The alert safety specialist will recognize these symptoms and report them to top management to correct the situation.

Before the supervisors receive any special training, it should first be determined what type and level of training is needed. It does more harm than good to go in with too high or too low a level of training or conduct training in areas where it is not needed.

Another major responsibility of a plant safety department is to become acquainted with the local fire department and fellow safety specialists in nearby industries. This type of relationship can be invaluable in a disaster. In many industrial areas mutual aid groups are formed. Each plant provides employees and equipment based on various formulas. The employees from all plants get together at least monthly for training. The training time is paid for by each employee's company. When an emergency arises the mutual aid group assembles to help the affected plant.

Part-Time Safety Specialists

The part-time safety function is probably the most difficult and frustrating of all safety functions. The employee is usually given the assignment when he already has more to do than he can get done in a normal day. More than likely the operation is small and the part-time man will be called on to function as a line safety specialist.

If the company has a corporate staff, they should lend as much help as possible to the part-time employee. Certainly he should not attempt to handle the companion jobs to safety, such as the insurance, medical, or security jobs. The part-time safety specialist is more dependent upon the help of all plant supervisory personnel than is the full-time specialist. He must also have the support of the local plant management in assigning various supervisors to develop policies and procedures, conduct inspections, and hold safety meetings and training sessions. If the local plant is not capable of developing its own policies and procedures and there is no corporate staff, a professional safety consultant should be engaged to help establish the safety program. After the program has gotten off to a good start the consultant should periodically review the operation to ensure that it is still headed in the right direction.

Some very good and lasting programs have been developed by part-time safety specialists. However, in most cases it is only fair to give them some kind of professional help in the early stages of the program.

There are many other ways that the part-time man can be upgraded. For instance, the National Safety Council has many courses such as the Key Man Course that are frequently scheduled in many different cities. The local safety council or state safety associations are a good place to get schedules of this training. Also, the National Safety Council has many courses in Chicago lasting one week or more that may be attended for a minimal fee.

The company workers' compensation carrier normally has a good safety staff that can provide information on implementing a good safety program. Usually these safety specialists are very good in their field. A major problem with using the insurance safety specialist is that, since OSHA, most of these people have such demands on their time, it is sometimes difficult for them to spend sufficient time with a company to establish a safety program. Regardless of whether consulting help is available or not, the part-time man should not get discouraged and give up. In many plants it is surprising to see how most employees will make extra effort to establish a safety program when they see that management is really serious about improving the job safety and health.

Safety Consultants

The last of the functions to discuss is the loss control or safety consultant. This is one group of safety specialists that can add new dimensions to any safety program. There are several points that anyone planning to hire a consultant should consider. They are (1) Is the individual or firm qualified to perform the duties you desire? (2) If it is a firm of several employees, are you assured of receiving the same quality of service from each person? Usually it is preferable to use only one individual from a firm to do the consulting unless a specialized problem develops. (3) Always check the reference of anyone that is to be considered as a consultant. If he cannot furnish references, it is risky to hire him. (4) Does the consultant have sufficient liability insurance for the type of work he is to perform? (5) The individual should have some certification such as Certified Safety Professional or equivalent qualifications.

When engaging the consultant a definite contract should be drawn. It should specify as a minimum, services desired, rates per hour, number of days after services that reports are due, and how many copies and to whom the copies are to be sent. If items such as these are established in the contract, usually a much better relationship will be maintained between the company and the consultant.

There are many types of consultants that may be engaged, such as safety, fire protection, industrial hygiene, security, and product safety. The fire protection engineer serves a very important function when developing fire protection plans for a facility. Fire protection engineering is a highly specialized field in which special engineering disciplines are utilized. The solution to many fire protection problems requires a combination of training and experience that can only be found in a fire protection engineer.

The safety manager in industries such as textiles, chemical, petroleum, and others with high risk fire potential should use the services of a fire protection engineer. This is especially true where new or modified facilities are planned. The fire protection engineer should be brought in as early as possible so that he may review plans and specifications for fire prevention and protection adequacy.

Anyone planning to consult a fire protection engineer should make sure the fire protection engineer has experience in the user's industry or at least in a related industry. This may not be necessary in all instances, but in most cases each type of industry has specialized problems that can best be solved by industry experience.

Companies with high risk fire potential, and which are large enough, should have a fire protection engineer on the corporate safety staff. When one is on the staff, continuous fire protection engineering surveillance can be conducted on all facilities. The use of the fire protection engineer on detailed fire prevention and protection inspections will usually pay large dividends. Regardless of whether the fire protection engineer is a company employee or a private consultant, the safety professional should use this type of expertise whenever possible.

Staying on Top of the Problems

In any safety program there will be times when it may appear that the program is falling apart. A series of accidents may begin to happen in areas which have been relatively accident free. If this happens immediate action should be taken. Far too many safety specialists do not recognize these problem areas soon enough or sit back in their dream world and say that a trend has developed. Don't wait for a trend. If people are getting injured something is wrong—dig in and find the problem. There could be various reasons for the problems: new work rules or new equipment has been introduced into the work area, a new supervisor may have been placed in charge, or there may be labor problems. The point is to get in quickly before the program completely falls apart. It will be much easier to turn the program around at this point than it will to wait for a year or so for a trend to develop. This type of action is the mark of any good safety or loss control program.

3
Safety
Assessment

When a particular plant or location's safety performance is below an acceptable level, positive action must be taken to determine the causes. It is usually a combination of causes that must be found if corrective action is to be effective. The particular approach taken here is the safety assessment and follow-up actions indicated from the assessment. The safety assessment is performed by

1. Taking a safety attitude survey of supervisors.
2. Reviewing all accident records available up to three years.
   (a) Categorize inquiries by types.
   (b) Assigning causes or responsibility to each accident.
   (c) Determine accidents that were caused by lack of personal protective equipment.
   (d) Determine number of accidents by job.
   (e) Determine number of accidents under each supervisor.

The safety assessment may be conducted by one man or a well-organized team. In most cases, a one man survey is preferred if the man taking the survey is well-qualified.

Attitude Survey

Taking an attitude survey is probably the best way to begin a safety assessment. The survey is conducted by individually talking to a large number of supervisors from the facility manager to the first-line supervisor. At least 60% of the supervisors should be surveyed in private. Listed below are the types of questions that should be directed at each individual:

1. What do you feel is your current safety program?
2. Who do you think is responsible for safety?
3. What would you do to improve safety at this facility?
4. What should top management do to improve safety at this location?
5. How does bad safety performance affect your job?
6. What have you done in the past six months to improve the safety of the people you supervise?
7. How often are safety meetings held?
8. Do you think your safety meetings are meaningful?
9. Who conducts the majority of your safety meetings?
10. How much authority do you have to correct unsafe conditions?
11. How much supervisory safety training have you received since, or just preceding being made a supervisor?
12. Who makes most of the safety decisions at this location?

When the predetermined supervisors have been surveyed, answers to each question should be tabulated and categorized. After this is completed, it should be fairly easy to determine which levels of management are deficient in carrying out their safety responsibility and accountability. This procedure can also determine the current safety problems that are considered most significant by the supervisors. It identifies the strength and weakness in safety training and safety meetings.

All accident records for the past three years, if available, should be reviewed. In these reviews, two steps should be taken.

Table 3-1. Injury Assessment		
Types of Injuries	**First Year**	**Second Year**
Strains	4	9
Bruises/Cuts	18	7
Broken Bones	4	0
Eye Injuries	8	1
Chemical Inhalation	4	0
Chemical Burns	0	3
Thermal Burns	6	4
Back Injuries	8	7
Deaths	0	1
Miscellaneous	1	0

The first step is to categorize all accidents by types as shown in the assessments in Table 3-1. This gives the person conducting the assessment the types of accidents that are occurring and the condition of the plant's physical facilities.

Accident Responsibility

The second step, and no doubt the most important step, is to assign responsibility and cause to each accident. In most cases two or three causes can easily be assigned to each accident case. These causes are management responsibility, supervisor responsibility, employee error, mechanical design, and mechanical failure. The idea here is to assign responsibility to each area that contributed to the accident. For instance, in the second year assessment in Table 3-2, 55 causes were assigned to 32 accidents. When the areas of responsibility have been determined, corrective actions may be begun.

Employee Error

Many employee errors can have several causes such as

1. Lack of training.
2. Poor judgment.
3. Poor health.
4. Poor attitude.

Table 3-2. Accident-Cause Correlation		
Cause	**First Year** (53 accidents)	**Second Year** (32 accidents)
Employee Error	43	27
Supervisor Responsibility	33	13
Management Responsibility	17	10
Mechanical Design	2	3
Mechanical Failure	1	2

These causes, as one would suspect, call for different corrective actions. Management must decide which of these causes caused most of the errors and commence the appropriate corrective action.

Management Indifference

Much supervisor responsibility can be traced to factors, such as

1. Lack of supervisory skills.
2. Lack of management support in established company policies.
3. Lack of company policies in discipline.
4. Supervisor unqualified for supervisory duties.

Lack of supervisory skills usually is caused by absence of a supervisory development program. In many cases an employee is tapped for a supervisor position and put to work without any preparation. Also, he has just been chosen over his fellow workers, who in many cases, feel they should have had the position, and this makes his task doubly hard as he will get only reluctant cooperation until he has proven himself to his workers.

Without presupervisory training, he will likely make errors that will haunt him for many of his supervisory years. He will continue to think and work much as he did before he became a supervisor. Over a period of time he will gradually develop some supervisory skills, but in most cases will lack the necessary skills required of most supervisors.

Lack of management support in discipline cases many times can be tied directly to the lack of supervisory skills. The supervisor has not been trained to handle discipline cases so when he bungles the job, management may be unable to support him because of labor contracts or other reasons. This makes the supervisor feel that he is in the middle and no one really cares about him, as long as the work is performed in a reasonable time and manner.

Lack of company policies on discipline causes much the same problems as lack of management support. Consequently, the supervisor bungles the job of discipline or does not discipline at all and has no control over his employees.

A supervisor unqualified for supervisory duties may be a controversial area. However, many employees are promoted into supervisory ranks simply by reason of seniority. Seniority alone does not qualify any employee to become a supervisor. His safety attitude has much to do with how he performs as a supervisor, so it should be reviewed just as his other qualifications for the job are reviewed. Poor safety performance should be an automatic disqualification.

Causes of accidents for which management is responsible are probably the easiest to cure once they have been pointed out. This is because most managers who are concerned enough to initiate a safety assessment have enough motivation to start corrective action. However, it must be pointed out that many of these accident causes require major financial expenditures. In many cases these causes have built up over the years because tightfisted management in the past would not spend the necessary money to keep the facility in a condition that is conducive to good safety performance.

As soon as management starts to let a facility deteriorate, poor production, quality, and safety will soon follow. If some reasonable explanation is made to the employees, such as bad economic conditions, slump in sales, or whatever the cause may be, the preceding statement may not be true. In fact, many responsive work forces will expend extra effort when a company has problems if they are only made aware of the problems.

Mechanical Failure

Mechanical failure causes in most assessments usually amount to less than 10% of the total causes. However, when a mechanical failure does occur, the results will usually be more severe than the other causes. When a mechanical failure occurs, a complete and thorough engineering review by competent engineers should be conducted. In many cases the mechanical failure review will show shortcomings in preventive maintenance. This, in turn, points out shortcomings in the management and supervisory ranks.

Mechanical Design

Mechanical design causes usually amount to two or three percent of the causes. However, an accident caused by bad or poor design can be catastrophic in many industries. As in a mechanical failure, a mechanical design cause, calls for an engineering review team to be assembled to make the required design changes.

Protective Equipment

It is necessary to determine accidents caused by lack of personal protective equipment. The accidents one looks for here are eye injuries, foot injuries, and other injuries to the musculoskeletal part of the body. If there are several eye injuries, then probably an effective eye protection program is required. It may be that safety glasses are all that is required, or it may be that full chemical goggles or face shields are needed. The kind of program needed can easily be determined as the need arises. Foot protection and protection for other parts of the body must also be determined as the need arises. It must be pointed out that governmental regulations now require that protection be worn where needed, or the company may be cited and possibly fined upon violation of these regulations.

Industrial Illness

Many individuals who are in charge of the company's safety effort fail to recognize industrial illnesses, which are extremely important. Some causes of industrial illnesses can be fairly simple to correct, or as the case in many plants, they can be extremely costly, even to the point of almost causing bankruptcy.

With the 1970 Occupational Safety and Health Act came a new awareness of industrial illnesses. The main thrust has been in hearing losses, carcinogenic agents, such as asbestos fibers, or carcinogenic gases. There seems to be little doubt that only the surface has been scratched in the area of industrial health.

It is imperative in today's world for anyone who is designing and building a new facility to make it as quiet as possible and to prevent the escape of fumes, gases, or dusts into the work or outside environment.

Accident determination by jobs should also be done. When this determination is made one should look for patterns in similar jobs. It may well be that a large number of accidents are being caused because a job is being performed in an unsafe manner. Sometimes only a slight adjustment is required to prevent injuries. Areas where changes may be needed are lighting, ventilating, heating or cooling, work speed, and whether or not there are crowded work conditions. Although this list can almost be endless, it can be very beneficial in reducing injuries. In fact

Table 3-3. Accidents Caused by Lack of Protective Equipment

Accident	First Year	Second Year
Eye Injuries	8	0
Chemical Inhalation	4	0
Chemical Burns	2	2
Thermal Burns	4	0
Foot Injuries	1	0
Deaths	0	1

Table 3-4. Incidence of Accidents by Job Title

Job Title	First Year	Second Year
Foreman	0	3
Secretary	0	1
Barrel Washer	0	1
Supervisor	0	2
Truck Driver	1	0
Insulator	1	0
Helper-Operations	6	0
Operator	4	3
Pipefitter	10	2
Janitor	1	0
Oil Canner	1	0
Welder	3	2
Instrument Man	3	0
Electrician	4	3
Boilermaker	2	1
Rigger	6	2
Machinist	1	1
Helper-Maintenance	2	0
Pump Repair	2	0
Laborer	1	1
Compounder	1	1
Tank Car Loader	1	0
Chilling Engineer	1	0
Truck Loader	2	0
Helper	0	2
Blender	0	1
Operator-Decoking Crew	0	1
Painter	0	1
Sample Tester	0	1
Mechanic	0	3
Gauger	0	1

on every accident one should consider these points as he performs the accident investigation.

A safety assessment was conducted in one location at the end of the first year. When the assessment showed the need for a supervisory training program, immediate steps were taken to institute one. The supervisory training program was not the only neglected area. However, it was thought to be the most serious and the necessary first step in bringing an effective safety program to this location.

Table 3-5. Accidents by Supervisors-Maintenance		
Supervisor	**First Year**	**Second Year**
H	10	4
E	11	4
C	3	3
K	3	2
R	3	0
L	3	3
S	1	0
T	3	0
N	0	1
B	0	1
U	9	0
V	1	0
Q	1	2
A	2	2
D	1	1
W	2	0
F	0	2
G	0	1
I	0	2
J	0	2
M	0	2
O	0	1

At the end of the second year a follow-up assessment was conducted to try and determine the effectiveness of the training. This training was completed about midyear.

The results were

1. OSHA reportables down about 30%.
2. The maintenance group, where intensified effort was invoked by the supervisors, had a 50% reduction in OSHA reportables.
3. The operations group, which exerted little effort beyond the class room, maintained about the same number of accidents.

For a complete breakdown of the accidents see Tables 3-1 to 3-6.

4
Determining the Causes of Accidents

To be successful in accident prevention one must realize that accidents are caused. As much concern should be shown *before* the accident as after it. It is more important for management to be concerned with what *can cause* an accident, than with what *has caused* an accident. By eliminating all detectable causes, the number of actual accidents will be greatly decreased. As management becomes more proficient in eliminating accident causes, safety performance will improve accordingly.

Potential and Actual Causes

Accident causes must be divided into two kinds, potential causes and actual causes. Potential causes are those which may be triggered at any time to cause an accident. Unsafe acts and unsafe conditions are potential causes because if allowed to continue they will cause an accident. Personal factors are also causes, such as, lack of skill, poor vision, or being under the influence of alcohol or drugs.

Actual causes are those which have developed into accidents. These causes are identified only after an accident.

Certain unsafe practices or unsafe acts may be a potential cause at one time and an actual cause at another time. When looking for potential causes one should

1. Check for unsafe practices.
2. Determine the reason for the personal factor causes.
3. Check for unsafe conditions.
4. Identify the source of the unsafe conditions.

Detecting and Eliminating Unsafe Practices and Conditions

The following two techniques are needed to discover unsafe practices:

1. Supervisors must be more than alert for unsafe practices. They must deliberately establish a procedure for spotting unsafe practices. This procedure should include the supervisor's observing a specific worker doing a specific hazardous job.
2. Job safety analysis is a method for analyzing job procedures in order to develop better and safer job procedures. The job is broken down into steps and each step is analyzed for hazards.

These next two procedures are used in eliminating the personal factor causes:

1. Supervisors must constantly be on the alert for personal factor causes that may cause a man to work unsafely. They should look for signs and symptoms which indicate a lack of job knowledge and skill. The supervisor should be aware that physical or mental conditions can cause an employee to work unsafely. Sometimes an employee's attitude may cause unsafe work habits. The supervisor should not wait for such personal factors to cause unsafe practices and accidents before attempting to stop them. Prompt action on

these signs and symptoms by the supervisor may prevent an unsafe practice that could cause a serious accident.

2. Repeated unsafe practices must be eliminated. The supervisor must correct each employee who is observed working unsafely. One should also determine what caused him or her to act unsafely. Different causes, such as lack of job skills, poor eyesight, improper safety attitude, or just honest ignorance, require different corrective action.

Unsafe conditions are continuously being created by worker actions, normal wear and tear, and errors in equipment or facility designs. Each new unsafe condition adds to the total picture and care must be taken to ensure that they are discovered and eliminated promptly. Early discovery of unsafe conditions is a basic principle of accident prevention. There are five procedures that may be used in detecting unsafe conditions:

1. Every operating area should have a supervisor's safety inspection program. Supervisors must know their assigned responsibilities and what conditions to look for in these inspections. Each supervisor should devise a check list so nothing will be overlooked.

2. Governmental regulations require that certain types of equipment, such as boilers or pressure vessels, be inspected by a licensed outside inspector. This is known as a specialist safety inspection.

3. It is the employee's responsibility as well as management's responsibility to report unsafe conditions. The employee can correct many of the unsafe conditions as he routinely performs his daily assigned task. Each employee should be assigned an inspection area.

4. Each operating unit should have a nonscheduled management inspection at least monthly. It has two main benefits: (a) it allows unsafe conditions that others have overlooked to be discovered, and (b) it serves as a motivator to bring about more effective inspection by supervisors.

5. Staff safety inspections should be done periodically to determine the effectiveness of the normal safety inspections.

Many employees think that only safety rule violations are unsafe acts. Unsafe acts are in most cases not safety rule violations as it is impossible to establish rules to cover all the ways an employee may find to act unsafely. All unsafe acts should be corrected as soon as they are observed. The following are types of unsafe acts that should be reviewed in training and safety meetings:

1. Occasionally, employees may use equipment or operate machines which they are not qualified to use and without authorization.
2. Equipment or material that is subject to automatic start up or unexpected movement such as slipping, sliding, rolling, falling, or drifting must be made secure against such actions.
3. Many workers will try to complete a job too quickly. They will operate a machine or vehicle at excessive speed. They may take shortcuts on a job in order to finish quicker. Rushing a job by running, throwing, improvising, or taking shortcuts increases the risk of an accident. Management must be careful not to encourage this type of activity by giving bonuses or other incentives for greater output.
4. If machinery is about to be started up or mobile equipment is about to be moved, a warning signal should be given by the operator. Also, temporary hazards such as floor openings, overhead work, or excavations, require the use of appropriate warning devices.
5. Most equipment is provided with some type of safety device. It is a very serious practice to make such devices inoperative by tampering with them, yet many workers do so. Specific examples include removing machine guards, disconnecting speed governors, making limits switches in-

operative, removing ground wires from power tools, and removing safety tags and locks before equipment is safe to operate.

6. Tools and equipment will develop defective conditions through normal wear and at times because of misuse or abuse. Naturally, constant checks should be made so that they may be repaired or replaced. Use of equipment in a defective condition runs the risk of an accident. Some examples of this type of unsafe practice are using tools with loose or cracked handles, chisels with mushroom heads, ladders with cracked rungs, cables that are kinked or frayed, ropes with broken strands, and vehicles or equipment with defective brakes.

7. It is not uncommon for employees to use sound tools and equipment in an unsafe way. Frequently the tool or equipment is used for a purpose other than that for which it was designed, such as a pair of pliers for a wrench. This may damage the bolt head as well as cause an injury. Sometimes the tool or equipment is used for the right purpose but in the wrong way.

8. Accidents may occur when workers place themselves in hazardous locations. Examples of this are working beneath suspended load, working beneath scaffolding which is not equipped with toe boards, or working near a traffic lane without guards or lights.

9. Repairing of equipment or service work should not be performed when the equipment is either moving, energized, or pressurized. This type of work requires the use of appropriate shut-downs, lock-outs, and start-up procedures.

10. Anything that can be ridden on is always a problem in terms of unauthorized riding. Trucks, tractors, fork lifts, motorized carts, crane hooks, conveyors, or any equipment that moves will attract riders. Many workers have been known to expose themselves to serious injuries to ride a short distance.

11. Many accidents with serious injury consequences are caused by horseplay. Horseplay must be recognized as a serious type of unsafe act. Bottle tossing, poking the unaware or jumpy worker, or similar tricks must not be allowed by the supervisor.

12. Many jobs require the use of personal protective equipment such as hard hats, safety glasses, or safety shoes. Past accident experience in the same or similar type of work may indicate the need for such equipment.

13. The wearing of unsafe clothing is also an unsafe act. Examples of this unsafe act include wearing oily or greasy clothing, wearing clothing that fits loosely or has strings or straps hanging loosely, the non-wearing of shirts, or the wearing of sneakers or sandals.

14. Another type of unsafe act is the body position a worker takes in relation to his work. Examples of an unsafe posture are lifting a heavy object with the legs straight and back arched down to the load, or trying to straddle an opening that is too wide.

An unsafe act does not usually mean an accident. Many unsafe acts are repeated numerous times without an accident occurring. However, the accident will catch up with the unsafe practice. The worker eventually will have an accident. It should be recognized that some unsafe acts will cause accidents much quicker than others, also some unsafe acts will result in much more serious consequences than others.

When a worker knows he is performing an unsafe practice and gets away with it, without an accident, he is developing a bad habit. He knows he got away with it and tends to lose his respect for the hazard. Each time he gets by, the worker becomes more convinced that the hazards of the unsafe practice are greatly exaggerated.

Once a worker has developed the habit of an unsafe practice, it is very difficult to correct because he does not believe the hazard exists. His experience of getting away with it so many times in the past makes him have a deaf ear to almost anything a supervisor may say about the hazard. A supervisor must never permit an unsafe practice to become a habit. He should always try to stop the practice by promptly correcting any unsafe practice observed.

The supervisor should always be alert for any unsafe conditions of tools, equipment, machinery, materials, structures, or other elements of the working environment that may cause or contribute to an accident. At times it may be impossible to correct an unsafe condition immediately. However, if proper analysis of the situation is made, work may be allowed to continue under stricter conditions. The operation, needless to say, should be brought back to its original safe working condition as soon as possible.

Examples of Unsafe Conditions

The supervisor should constantly be on the alert for the following types of unsafe conditions:

1. The absence or inadequacy of machine guards and other safety devices where they are needed are unsafe conditions.

2. Certain work situations require the use of effective warning devices to warn workmen of existing or potential hazards. Warning devices are needed on unguarded floor openings, overhead work, mobile equipment, automatic starting equipment, and hazardous work operations that require areas to be kept clear. It is very important to use the proper warning device for the job, taking into consideration such things as background noise or visibility.

3. Storage of combustibles or explosives in unauthorized areas or in ways that create fire or explosion hazards are unsafe conditions.

4. Sometimes machines, vehicles, and other types of equipment as well as materials are left in a position where they

may roll, slide, or topple over. Proper procedures should be followed which prevent this type of unsafe condition.

5. Tripping and slipping accidents are indicative of poor housekeeping. Tripping hazards exist whenever tools, debris, and other obstacles to foot movement are in areas of pedestrian traffic. Slipping hazards exist when water, and other traction robbing materials are in areas of pedestrian traffic. Steps and other walking areas also should be designed to prevent slipping hazards. It is important for the designer to realize that the safety of many workers may depend on the materials he specifies for a job, or the surface finish he specifies.

6. Many workmen are injured when they are caught on objects that protrude into their working and walking areas. If objects that cannot be eliminated protrude into the work area a guard or warning device should be installed.

7. Close clearance and congestion are usually unsafe conditions as they are likely to cause accidents. Sometimes a guard may be installed, in other cases equipment or supplies should be relocated.

8. Hazardous conditions exist whenever the air workers breathe contains harmful concentrations of toxic gases, injurious airborne chemicals, particles or dust, or a deficiency of oxygen. This is one of the most severe, yet least recognized, of all unsafe conditions in many work environments.

9. Improper storage and arrangement of supplies, finished products, and equipment used in the work environment sometimes create unsafe conditions.

10. Tools, equipment, machines, and structures may become hazardous from use, misuse, and abuse. Specific examples include cracked ladder rungs, rotted ropes, cables with kinks and broken strands, frayed electrical cables, and corroded piping.

11. The lighting in the work environment may be a source of accidents or industrial illnesses such as headaches or eye strain. This does not mean that the light level is too low, in fact it may be too bright, which causes glare, or there may be shadows from improper location of light sources.
12. Wearing personal attire that creates an unsafe condition should not be allowed. Clothing policies for each work area should be developed and used.

Many unsafe acts and conditions may be identified in a supervisor's inspection. However, if the supervisor fails to follow up to ensure correction he has wasted his time. When one of these acts or conditions is discovered, it should be written down. Action should be taken to start the wheels moving to correct the situation. At least once each week this list should be checked to see if the necessary action has been performed. If not, give an extra push and stay with it until the necessary corrective action has been performed.

5
Safety
Meetings

Safety meetings are very important in improving safety performance. To achieve and improve safety performance from the safety meetings, the purpose of the meeting must be fully defined and rigidly adhered to. Never let a safety meeting become a gripe or joke session.

The purpose of any safety meeting is to inform the group on how to perform a specific duty safely, to pass on safety information, and to establish a safe work environment in the particular plant or location. It should be a planned meeting, where adequate time and effort are spent in preparation for the meeting. There are several types of safety meetings, preparation and presentation for each is somewhat different.

Personal Safety Contacts

It is neither necessary nor desirable to conduct all personal safety contacts in the same way. There should be variations in method as well as subject matter. Four methods that may be used are

1. Explain to the employee that you are interested in hearing how he thinks a particular job or part of a job should be

done safely. Ask him to explain or demonstrate how he does it. If he is properly trained he will give you a good account. If he omits any important precaution try to draw it out of him with a tactful question. The idea is to draw out of him all the major precautions related to the job or step. You'll find most employees eager to explain what they know. If you give proper attention to what the employee has said you will know very quickly if he has the right safety attitude and if he knows how to perform the job safely.

2. Have one employee explain a particular job to a co-worker. Arouse his willingness by letting him know that you think it would help if the employee heard it coming from him. Get them together and explain the situation to the employee. Stand by as a neutral party, and only interrupt if an important point is omitted. If you must interrupt it is very important to use a tactful and leading question that will result in the desired answer. The most important thing about this approach is to avoid saying anything that might embarrass the experienced employee. Never flatly contradict him. If you handled it right one employee will walk away feeling satisfied, and maybe even feeling a little proud of himself, and the other will have learned an important safety lesson. Two safety contacts will have been made very easily.

3. If you have a specialized job under your supervision that you know less about than the people doing it, ask them to explain the safety hazards and necessary precautions to you. Show that you really are interested in the safety problems related to the particular job. Usually a worker will be happy to discuss the safety aspects of the job.

4. To emphasize unsafe practices try this approach. Tell the employee you are collecting information on accidents caused by a certain unsafe practice. Discuss the unsafe practice and ask him if he knows of any accident resulting from such an unsafe practice. If he doesn't know one, have

one handy to repeat to him. Always handle the conversation so that you remind him of a safety rule or a safe practice.

Most supervisors don't like to make formal sounding safety talks, which is just as well because personal safety contacts are better than safety talks. They are easy to make and take very little preparation. They accomplish the purpose of a safety talk by reminding employees about hazards, safety rules, unsafe practices, and therefore help shape safety minded attitudes.

Weekly Safety Meetings

This type of meeting has many names depending on the industry where it is used. It should be held at the beginning of the shift on the first work day of each week. This meeting should be limited to one subject that can be discussed in five minutes. It should emphasize safety and establish a safe working attitude for the work week to follow. When this type of approach to safety is used, employees get into the safety habit by starting the week with safety being the first thing emphasized. This establishes the company philosophy that lets all employees know that management insists on safe production.

Subjects that can be made into five-minute safety meetings are

1. Recent accidents, that have occurred in a similar type work environment.
2. Unsafe practices or work conditions observed by the supervisor make an excellent subject. These subjects are particularly good if they have been on the increase. This is a very effective way to discuss the subject quickly without identifying any individual who might be embarrassed if he were confronted person-to-person about practices or conditions. This is not to say that an employee should not be corrected if he is observed working unsafely or creating unsafe conditions. However, as a supervisor becomes acquainted with his employees' individual personalities he can make a

rational judgment as to which method he should use to correct the unsafe practice and condition in his area of supervision.

Group Safety Meetings

The group safety meeting is often the most effective way to conduct regular safety meetings in many operations. This type of meeting is usually held no more than once a month. In practice there is only one problem with this type of safety meeting; that is, very few supervisors have developed the necessary skills to conduct interesting and effective safety meetings. One main reason for this lack of skill is that very little has been written that actually gives guidance on how to conduct safety meetings. Usually, such meetings are poorly planned and badly conducted. If this weakness is recognized and someone thinks a cure is necessary, the "canned" safety talk is a frequently tried remedy. The supervisor is told to find a source, or give a printed version of safety talks, and use it on his group. The subject is usually related to his operation, and at best in a general way. Sometimes it isn't related to his operation at all. The net result is that he is sidetracked from the kind of subjects that he should be discussing with his employees. In many cases a canned talk is not any better than what had been used in the past.

Group safety meetings should be used in the following circumstances:

1. The subject needs to be covered promptly with a large number of employees.
2. The subject is especially suitable for a lively group discussion.
3. The subject involves a demonstration.

The following subjects make good group safety meetings:

1. Safety rules and regulations that apply to most employees, such as the wearing of safety glasses or hard hats, are especially suited for this type of meeting. They should be

used when they are newly introduced, or involve increased safety violations.

2. Safe job procedures make a good safety subject. The group situation permits a sharing of common experiences. When instruction of the subject is improved by a demonstration, such as how to use a fire extinguisher, a group meeting is the only practical way. Personal demonstrations make sense only when a subject is related to one or two employees' jobs.

3. Group meetings are appropriate when higher management wants a particular subject discussed with all employees. For example, this type of subject could be employees' safety responsibility.

Special Safety Meetings

The special safety meeting is much like the group safety meeting. The major difference is that it is an unscheduled meeting. When it is used, it brings special attention to the subject at hand, thus allowing management to show its feelings about the subject by allotting extra time to cover it. Instances when a special safety meeting should be called are

1. When a major expansion or maintenance project is begun.
2. Just prior to the start-up of major pieces of equipment involving a large group of employees.
3. Shortly after an unusual or major accident where employees involved in the particular type of work are brought in to explain what is known about the accident. Any extra precautions that may have been pointed out by the accident should be covered.

Proper planning and selection of a strong safety subject for the scheduled meeting is of upmost importance. New life can be added to the safety meetings when this approach is used. Safety meetings just as any other meeting conducted by management should be planned and guided in the direction of the immediate need. Efficient management does not call employees together just to be a gathering of people under the guise of a production, budget, or any other meeting. So, if safety is to find its proper place in the management system, it must be more than a gathering of employees.

Selecting a Safety Subject

Select a subject that will be of interest to most employees at the meeting. If a subject does not apply to most of the employees, but still needs to be discussed with part of the group, the group should be split and appropriate subjects selected for each group. Have a good reason for selecting a particular subject. The following are some good reasons:

1. The subject is directed at trying to change attitudes.
2. The subject has a great potential for preventing serious injury or large financial losses.
3. The subject is of such nature that employees need to be periodically refreshed in its potential hazards.
4. The subject has been recently reviewed by management and a change in the recommended safety procedure must be initiated.
5. The subject is new, such as a new fire truck or a fixed fire fighting system, and its operation must be known to all employees.

If the subject is of such nature that it can be covered by job safety analysis, it should be reviewed in the early planning stages. If a job safety analysis is not available, develop one to ensure that all hazards and steps of the job are covered. Whatever the subject matter is that you are to discuss, make sure that you can intelligently discuss all hazards and have reasonable safeguards against the hazards.

During the meeting, it is very important to arouse interest when you introduce a subject. If the subject has been properly selected, telling the group why it was selected will usually be all that is needed. Once a subject has been introduced be prepared to get right into your discussion. If questions are asked, by all means answer them and try for audience participation. However, stick to the subject and don't be drawn off into unrelated areas. At the end of every meeting safety suggestions or hazards should be received and discussed by the group.

All group meetings require that minutes be recorded. These minutes should contain

1. A brief summary of the subject covered.
2. A list of all safety suggestions received.
3. Safety suggestions that were disposed of in the meeting and their dispositon, i.e., by corrective action, by no action, or by consulting management.
4. Comments to the next level of management about how the suggestions were treated.

Planned safety meetings conducted on a regular basis, are the most effective ways of educating people about the hazards of their work, applicable safety rules and regulations that govern their work, and safe job procedures in general. Such meetings do not have to be a part of department wide or plant wide program to be effective. Nevertheless, the maximum benefits are obtained when all supervisors conduct such meetings regularly.

The effectiveness of a planned safety meeting depends largely on how well supervision and top management stimulate, guide, and control the program. You can be sure any safety meeting program will bog down if your upper supervision takes no active role. Usually the meetings will not be held or they will be only token gestures, poorly prepared, badly executed, and useless to the supervisor and workman alike. As with other accident prevention tools, planned safety meetings require the active support and guidance of higher supervision for maximum results.

The weakest link in many safety programs is the scheduled safety meeting. A safety meeting that is just a gathering of people without planned direction does not normally obtain the desired results. Plan your safety meetings so that you demand action and obtain results.

6
Accident
Investigation

Many first-line supervisors, and also some high-level supervisors, have only a vague idea of accident investigation. They tend to think that it is merely a matter of getting and reporting a workman's account of his accident. This is a very poor way to think about an accident investigation. It reduces the investigation to a mechanical routine that could be performed by any office clerk with a check list. The first-line supervisor who is actually in charge of the accident victim must be held accountable for the accident investigation.

An accident investigation is a systematic effort to establish all relevant facts and interpretations of "how" and "why" an accident occurred, so that conclusions may be drawn about what must be done to prevent recurrence. Preventing recurrence is the true objective of the accident investigation, and every accident investigator should keep this in mind. When an accident occurs, the investigator should list all known possible causes. Each cause that cannot be eliminated after a thorough investigation, should be considered a contributing cause of the accident. It is doubtful that very many accidents will have only one cause.

Accidents should be investigated as soon after their occurrence as possible. The more time that is allowed to lapse before questioning witnesses and appraising the scene of the accident, the greater the risk of getting an inaccurate account of what happened and why. People forget accident details very quickly, particularly on the impact of emotional shock. An injured person's memory of an accident can easily become distorted. Also, witnesses, if allowed enough time, can exchange stories and thereby influence one another's version of what happened. People also imagine details that, in fact, did not occur. Probably the most important tool for use in early accident investigation is a good 35 mm camera. The camera should be capable of taking close-ups as well as normal pictures. Be sure to get pictures of every possible angle of the accident. Color photos are much superior to black and white. The camera records the accident just as it happened, provided nothing except the injured is moved.

There are two circumstances under which it will be necessary to postpone questioning of the injured person:

1. If the questioning will cause delay in obtaining medical treatment for the injured. Prompt medical treatment is always the first step even on minor injuries. Most workers will resent questioning under such circumstances. Start your investigation in this instance by questioning witnesses or by examining the physical damage of equipment if any exists.

2. Never question an injured worker if he is upset or in pain, even though it may not delay medical attention. Under this circumstance, devote your efforts to reassurance and making the person comfortable until he can be given medical attention. If there is doubt about questioning the worker immediately after his accident, ask him if he would like to talk about it or if he would prefer to wait until later.

The Role of the First-Line Supervisor

It is very important that first-line supervisors be responsible for investigating accidents that occur under their supervision.

They are the most qualified to investigate accidents because of their constant contact with jobs, working conditions, and workers. They know most of the details of the jobs, procedures, hazards, environmental conditions, and any unusual circumstances. They should also know their employees' job experience and personal characteristics. Such information does not always ensure that first-line supervisors will make good accident investigators. However, it does provide the background necessary for a good investigation. Figure 6-1 shows a supervisor's accident investigation report which, when completed, gives a full description of an accident.

When first-line supervisors are required to investigate accidents that occur under their supervision, they can more readily see their responsibility for accident prevention. Many companies delegate clerks or safety personnel to complete accident reports and still have first-line supervisors investigate such accidents. This only tends to undermine a supervisor's sense of responsibility for accidents and will result in a poor investigation.

Any thorough accident investigation will expose to the investigator previously undetected hazards that can cause accidents. First-line supervisors, of all employees, need this information. They are responsible for training new people, checking for unsafe practices, looking for unsafe conditions. They must remind men about hazards, and act to prevent accidents. By investigating accidents, they will become more proficient in accident prevention.

In cases of major injuries, large fires and/or explosions, or accidents that are significant for some reason, an accident investigation team should be assembled. The team should include the immediate supervisor plus three to five members of management. It should include the person in charge of safety for the location where the accident occurred. In some investigations it may be desirable to bring in experts from other parts of the company. This may help get a clearer unbiased opinion of the causes of the accident. It also may help in getting executive approval and support if a large sum of money is required to eliminate the causes of the accident.

Supervisor's Accident Investigation Report

1. Name of injured _____ 2. Age _____
3. Sex _____ 4. Years of service _____
5. Time in present job _____ 6. Title/Job _____
7. Dept. _____ 8. Date of accident _____ 9. Time _____
10. Date and time medical attention received _____
11. Did employee go to doctor? _____ 12. Had you instructed employee regarding hazards of this job? _____
13. Severity of injury ( ) First aid ( ) Reportable ( ) Lost time ( ) Fatality
14. Exact location of accident _____
15. Nature of injury _____
16. Part of body affected _____
17. Detailed narrative description. (How did accident occur, why, objects, equipment, tools used, circumstances, assigned duties. Be specific.) _____
_____
18. Was weather a factor? _____ If yes, how? _____
19. Unsafe act by injured and/or others contributing to the accident. _____
_____
20. Unsafe mechanical/physical/environmental condition at time of accident. _____
_____
21. Unsafe personal factors (improper attitude, lack of knowledge or skill, etc.) _____
_____
22. What personal protective equipment is required for job? _____ Was equipment used? _____
23. What will be done to prevent recurrence of this type accident? _____
_____
24. What has been done to prevent recurrence of this type of accident? _____
_____
25. What were the contributing causes of this accident? _____
_____
26. When did you visit site of accident? Date _____
Time _____
27. Witnesses _____ _____ _____
28. Date prepared _____
29. Signature of foreman/supervisor _____
30. Title _____ 31. Department _____

Figure 6-1. An accident investigation report such as this ensures that all the necessary information is obtained.

**Next Level of Supervision's
Appraisal and Recommendation**

31. Do you agree with the results of this investigation? _____
    If no, why not? _____
    _____

32. What is probability of recurrence? _____
    _____

33. What should be done to prevent recurrence? _____
    _____

34. What have you done to prevent recurrence? _____
    _____

35. Estimate of property damage, if any. _____
    Date _____ Signature _____ Title _____

Figure 6-1. Continued.

Usually corrective actions to prevent recurrence of accidents
are implemented by the first-line supervisor. The prevention
implemented may be a new or reemphasized safe job procedure,
or elimination of the unsafe condition and its source. When
supervisors don't investigate the accident, many of the necessary corrective actions don't occur. This is especially true in the
case of minor injury or equipment damage accidents.

Getting the Facts

The main problem a supervisor encounters when interviewing
an accident victim is getting that person to tell the truth and
provide complete facts about what happened. There are several
reasons why employees could be reluctant to tell the truth. In
many cases the injured knows he committed a stupid act or
violated a basic safety rule. He is well aware that this could
cause an official reprimand or ridicule from fellow employees.
Some uninformed employees may fear the truth will jeopardize
the right to injury compensation. The actual reasons take many
forms, but the basic reason is fear of some kind.

The investigator must minimize fears that repress the truth
by acting in a manner that makes the individual willing to
cooperate. If the investigator says or does anything which inten-
sifies fear in any manner, he probably will not get truthful
answers. If he has a reputation of giving employees a hard time

in an accident investigation, workers may distort the facts. However, the cooperation problem is solved when a supervisor has good personal relations with his people. Mutual respect and good will usually will encourage honesty from the worker.

When interviewing a worker who has had an accident, explain that the purpose of an accident investigation is to learn what occurred, how it occurred, and why it occurred, so that future accidents of this type can be prevented. Reassure him that the purpose is to find facts and not place fault. Let him know that you need his cooperation. The object is to set the worker at ease about telling the truth, and also, to convince him that he is helping to prevent accidents by cooperating. The more you can personalize your introduction, the easier it will be to get him to cooperate.

Ask the person to explain what he was doing, how he was doing it, and what happened. If possible, take him to the scene of the accident. He can demonstrate his exact location, and point things out that might otherwise be difficult to explain. It is very difficult for many people to express themselves unless they can point or in some way relate their explanation to things around them.

The worker may want to show you what he was doing when the accident occurred. If so, ask him what he intends to demonstrate. Make certain that he does not demonstrate an unsafe practice. Have him proceed slowly, making sure he does not repeat the action that caused the accident.

The person should be allowed to explain his version without interruption. Interruptions have a tendency to draw some people off their thoughts, which may make the person change his story somewhat. Also, don't make any remarks which might antagonize the worker or put him on the defensive.

If the person's story of the accident leaves some points unclear, ask specific questions to obtain the needed information. At this point, limit your questions to establishing what he was doing, how he was doing it, and what happened. Save for later the questions relating to why he did what he did. Questions of this kind are more likely to arouse resistance. They should be asked only after details of what he did and what happened have been established.

You must question him in a friendly way. Don't ask questions in a way which may antagonize him. Don't ask leading questions, because the power of suggestion might elicit false answers. Your questions should be confined to the ones that will help you understand what happened.

Understanding the worker's explanation is of utmost importance. Make sure your interpretation is correct by describing the accident slowly. Check with him on key points so that he may correct any misunderstandings.

If his version appears contradictory, don't try to corner or trap him. Many people have difficulty explaining things exactly as they happen. Contradictions are usually an indication of this difficulty rather than evasiveness. If his version is not consistent with what you know about the job, leave him a face-saving out. Explain that you don't understand how the accident could have occurred in that manner, and ask if you have missed something. Ask him to go over the inconsistent points again.

The best way to bring the interview to a close is to discuss how to prevent recurrence of the accident. Place special emphasis on the precautions that will prevent recurrence by either repeating them, or asking him to state them. This is also a good time to ask him if he has any ideas that would help to make the job safer. If he does, they should be discussed. If an idea needs further developing, ask him to work on it and let you know the results. Always, if possible, end the interview on a friendly note.

Witnesses

Witnesses are a vital source of information in most investigations. In fatal injuries, a witness may be the only source of information the investigator has. In serious injuries, a witness may be the only means of verifying the account given by the injured. On minor injuries, a witness may be able to clarify some of the circumstances surrounding the accident better than the injured person.

Obviously, a witness is anyone who sees an accident. But, a witness can also be anyone who has knowledge related to the accident, even though he did not actually see it happen. This type of witness, known as an indirect witness, is often overlooked by investigators. Investigators usually begin by asking if anyone saw what happened, but it may be more helpful to ask if there is anyone who might understand how or why the accident occurred.

Witnesses present special problems. They may not want to reveal what they know if they think it will reflect badly on a co-worker. Under such circumstances, many employees will deny any knowledge of the accident. Others will withhold or distort the facts to try to protect a friend. Convincing witnesses to cooperate by giving a true and complete account of what they know is sometimes a problem. Witnesses should be interviewed much in the same manner as the person who has had the accident. They need to be reassured and encouraged. Reassure them of the investigation's purpose. Let them know that the full story may help prevent serious injury to another person. If it is consistent with company policy, let them know that their testimony will not result in disciplinary action for the injured person. Also, let them know that their testimony will in no way affect workman's compensation for the injured.

Interview witnesses promptly and separately. Some investigators get in a hurry and try to interview all witnesses at one time. This is a mistake. They are likely to influence each other's testimony in a group situation. Some will not tell what they know in front of co-workers. The investigators should interview the witnesses as quickly as possible to prevent distorted information.

If possible, a witness should be interviewed at the scene of the accident. Ask him to tell you all he saw or knows. Let the witness give his version with a minimum of interruptions. Limit your interruptions to requests for clarifications of something he said.

After getting his version, specific questions should be asked to clarify unclear points or to develop more or new information. Don't ask leading questions. Your questions should be phrased so that they are neutral. For example, it is better to ask, Was there a mechanical failure? instead of, Was the accident caused by a broken gear? It is very important to maintain a friendly attitude during questioning periods. Some witnesses are likely to regard lengthy questioning as an attempt to discredit their testimony. Always check with the witness to ensure your understanding of what he said. Let him know you want to be corrected if your understanding is not what he intended to convey.

A useful way to obtain information is to reenact the accident. Accidents are reenacted by having someone, usually the one who had the accident, demonstrate the circumstances leading to the accident. The person demonstrating the accident should assume the actual location and position that was taken at the time of the accident. Within limits, the worker should demonstrate what was being done and how. Accidents should be reenacted only when

1. Additional information is needed which cannot be obtained in any other manner.
2. It will help you check out an idea that will prevent recurrence of the accident.
3. It is the only way to verify a statement made by a witness or the accident victim.

Reenactment is usually not necessary in investigating most accidents and should not be attempted. Accidents have occurred during the reenactment of other accidents when proper precautions were not observed. Certain precautions should always be taken when reenacting accidents:

1. Make sure the person will not repeat the unsafe practice that contributed to the actual accident. Stop the employee before he takes any unnecessary risks or acts unsafely.

2. Have the worker demonstrate slowly. Make sure he explains each step as he demonstrates.
3. Never use an emotionally upset person to reenact an accident. If the individual is nervous, tense, or agitated, postpone reenactment until he is calm enough to proceed. Even then, watch for signs that he is becoming upset; if this occurs, the reenactment should be cancelled or done with someone not involved in the accident.
4. Co-workers should not be allowed to observe the reenactment. Onlookers only add confusion and risk to the situation. Exceptions can be made if training for other employees can be obtained and the person performing the reenactment does not object.

Investigation Questions

The purpose of an accident investigation is to prevent recurrence of the accident. The investigation must establish the kind of information that provides ideas for preventing recurrence. The following questions provide the basic information that every accident investigator should try to establish:

1. Who experienced the accident? The person experiencing the accident must be identified by full name and social security number. Such identification may result in preventing accident recurrence, because the name alone may suggest other kinds of information that should be developed, for example: (a) Has he had a similar accident before? (b) Has he had a lot of accidents in the recent past? (c) Was he under the influence of illegal drugs or alcohol? (d) Did his physical condition contribute to the accident? (e) Has he been working excessive overtime? These questions may prompt actions that will head off further accidents. Investigators should remember they are not investigating the accident of a check number, but that of a person. Learn as much about the person as possible because it can have considerable bearing on why the accident occurred.

2. When did the accident occur? Three kinds of information are wanted; the date of the accident, the exact time of the accident, and how long into the shift. This information is useful in a number of ways; it establishes the accident in relation to the potential witnesses, it identifies the accident in relation to the shift, peak production periods, weather conditions, it identifies the supervisor in charge, and it helps correlate other factors which may determine the cause of the accident. Times and dates may not mean anything to the investigator, but they may suggest ideas to others who read the report.

3. Where did the accident occur? There are many reasons why the location of the accident is important to the accident report. It permits others to go to the scene if required. Statistical analysis may be developed to establish the repetitiveness of accident locations. Analysis of this type is useful for accidents involving vehicle ways, walkways, traveling mobile equipment, and certain types of environmental conditions. Questions about the safety of the worker's location relative to the hazards around him may be explored.

4. What position was being worked? The position that the man was working at the time of the accident should always be identified on the accident report. Such information is necessary for statistical purposes. It makes possible a tabulation of accidents occurring in each position, thus enabling studies to be made of the accidents associated with various positions. Also, questions may be raised about a worker's experience, qualifications, and physical fitness to work in a particular position.

5. What job was being done? The job being performed at the time of the accident should always be identified on the accident report. *The job being done should not be confused with the position worked.* Job refers only to the specific task the man was performing when the accident occurred.

6. What occurred? The answer to this question is an exact account of the events that led up to and ended with the acci-

dent. Such a description should involve the necessary background information to set the stage for the sequence of events that caused the accident, the person's position with regard to his work environment (this should describe where the person was in relationship to those parts of his work environment that had something to do with the accident), how he was doing what he was doing (this should describe the essential details of the person's actions immediately prior to the accident), and what finally triggered the accident. Usually, but not always, something unexpected happens that actually triggers the accident. Here a description of the type of accident that occurred, such as "struck by ..." and "caught between ..." should be provided.

Not all accidents require answers to all six questions. Sometimes background information is not necessary. The exact position may be irrelevant, as in the case of a worker twisting his ankle while running. The worker may not be doing a job, such as a person slipping while taking a shower. Sometimes there is no triggering incident to describe, as in the case of a worker developing a hernia while lifting.

7. What were the accident causes? The investigator should remember that very few accidents have only one cause and should not stop the investigation until all causes are determined. A complete description of the causes should be obtained by asking (a) What did the worker who had the accident do, or fail to do, that contributed to the accident? The answer should be a specific description of what he did that caused the accident. Often there are several things a worker has done to cause his accident. As an example, an employee might use a wrench with a handle that is too short, use an oversized extension to obtain excess leverage, and stand on an uneven work surface. (b) What caused or influenced him to act as he did? (c) Was he taking short cuts to save time or effort? (d) Was he aware of the hazard? (e) Did he have sufficient job experience? (f) Was he under the

influence of intoxicants, illness, or emotional stress? Such information is useful because it frequently helps determine the type of corrective action that must be taken. (g) What defects or otherwise unsafe conditions of tools, equipment, machines, work area, contributed to the accident? (h) Did such conditions exist prior to the start of the job or did they develop during the course of the work? These conditions may have been the results of the man's own doing, as in the case of a man falling victim to his own poor housekeeping. Contributing conditions usually have a direct relationship to the accident, as in the case of a person slipping on a slick surface. However, sometimes the contributing conditions are only indirectly related to the accident. For example, a frozen valve is broken off by a worker using an extension. Poor lubrication would be a contributing condition if it caused the valve to freeze. Another example, a workman sees a five-gallon can of gasoline, and decides to pour some on an open fire to increase burning rate. The can explodes in his hands, and he is killed. Because the gasoline was accessible to unauthorized use, its poor storage could be a contributing condition.

Often, the sources of unsafe conditions that result in accidents are of a repetitive or long standing nature. As a result, any corrective action must be designed to withstand the test of time.

An accident investigation is not complete until it has developed the corrective actions necessary to prevent recurrence. Such corrective actions are usually implied by the direct and indirect causes of the accident. If there are several causes, there will be several kinds of corrective action necessary to prevent recurrence.

An accident investigation report should always make a distinction between

1. Corrective actions already taken at the time of report.
2. Corrective actions planned but not yet taken or ordered.
3. Corrective actions that are recommended.

An accident investigation should also have a place for the next level of supervision to agree or disagree with the findings of the investigation. This has two main purposes, to ensure that the investigator has performed an adequate investigation, and, if additional authority is needed, to obtain the necessary finances for corrective action.

7
Facility
Inspection

As discussed in an earlier chapter, successful accident prevention depends on discovering the causes of accidents. This chapter deals strictly with the types of inspections that are required to ensure that the facility is physically sound.

Fire Inspection

Due to the extreme consequences that may result from a fire, it is always prudent to perform fire inspections on a regular basis. When an inspection is made, equal attention should be paid to fire fighting equipment as to fire hazards. Too many fire inspections are "once-overs" and do not involve thorough inspection of each piece of equipment.

Fire fighting equipment generally falls into five definite types; the hand portable extinguishers, the wheel extinguishers, automatic or manual sprinklers, fixed chemical systems, and hose lines; and in many cases, a sixth type such as a company-owned fire truck. Each type of equipment takes a specific type and frequency of inspection.

Each type of equipment should have a check list. The person who is to perform these inspections must realize the importance of his inspection functions. There is nothing more dangerous

Hand Portable and Wheeled Unit Check List						
No.	Location	Description and Size	Fully Charged	Operable	Sealed	Comments
1	Lab	CO_2 15#	Yes	Yes	Yes	—

Date _____   Signed _____

Figure 7-1. A monthly check list simplifies the task of inspecting hand portable and wheeled fire extinguishers.

than a piece of fire fighting equipment that is checked off as in good working order when actually it will not operate. Training the inspector to fully recognize inoperative equipment involves more than handing him a check list and a five-minute lecture. The inspector should be familiar with all the components of each piece of equipment he is to inspect. Figures 7-1, 7-2, 7-3, 7-4, and 7-5 are typical check lists for each of the types of fire fighting equipment mentioned.

Each extinguisher should have a permanent history on file, and any maintenance or testing should be recorded. If the extinguisher is thought to be damaged, it should be hydrostatically tested immediately.

Fire Hazards

After the fire fighting equipment check is completed the next obvious part of the facility inspection is to check for fire hazards. Particular attention should be paid to improper storage of flammable liquid. It is not uncommon to find flammable liquids in hazardous areas because proper storage is not available.

A facility can very easily become piled with combustible materials such as paper, packing materials, and scrap lumber. Precaution must be made to make sure that unnecessary amounts are removed to proper storage. Because some raw

Automatic and Manual Water Sprinkler Check List

Sprinklers and Controls	No.	Location	Condition
Outside Sprinkler Valves			
Inside Sprinkler Valves			
Manual Sprinkler Control			

1. Reason for any closed automatic control valves ⸺⸺⸺
   ⸺⸺⸺⸺⸺⸺⸺⸺⸺⸺⸺⸺⸺⸺⸺

2. Were closed valves reopened and sealed? ⸺⸺⸺⸺
3. Was flow test conducted after opening closed valve? ⸺⸺⸺
4. Was flow check performed on manual system? ⸺⸺⸺
5. Were all heads open? ⸺⸺⸺⸺⸺⸺⸺⸺⸺⸺
6. Were plugged heads corrected? ⸺⸺⸺ , if not why? ⸺⸺
   ⸺⸺⸺⸺⸺⸺⸺⸺⸺⸺⸺⸺⸺⸺⸺

7. Dry pipe or preaction systems
   No. 1. Valve room heated ⸺⸺⸺ . Air pressure ⸺⸺#.
            Alarms tested and operative
8. Automatic sprinkler—check for rusted, corroded or painted heads. Will stock or building modification block spray pattern? ⸺⸺⸺⸺⸺⸺⸺⸺⸺⸺⸺
   Is any part of system exposed to freezing weather? ⸺⸺⸺

   Date ⸺⸺⸺⸺⸺⸺⸺   Signed ⸺⸺⸺⸺⸺

Figure 7-2. *The manual and automatic water sprinkler inspection information can be kept on the same check list.*

materials are combustible, it is wise to never have more than a two-day supply of raw material and a one-day supply of finished products in the manufacturing or process area. It is very important to move the more valuable finished product into the proper staging or shipping area. Ensure that aisles are not blocked by any material as this is both a fire and safety hazard.

Smoking is a continous problem in most facilities, and should only be allowed in designated areas. Currently nonsmokers are

---

Check List for Fixed Chemical Systems

Monthly

System No.	Discharge Heads	Visible Physical Damage	Initials	Date

Semiannually

System No.	Pressure or Weight of Expellant	Dry Chemical *Only* Weight of Chemical	Initials	Date

Annually

Dry Chemicals Systems—except stored pressure checked for energy or packed chemicals at top center and near the wall. Halon Systems—thoroughly inspect and test for proper operation.

Figure 7-3. *Fixed chemical systems (halon and dry chemicals) are subject to monthly, semiannual, and annual inspections.*

demanding that smokers be prohibited from smoking in areas where nonsmokers must assemble. Specific note should be made of smoking in unauthorized areas. It is not necessary to actually see someone smoking, as careful observation can detect signs such as cigarette butts or ashes.

Environmental Inspection

The environmental condition of the facility is another important area of inspection. Dusty conditions may exist which can create health problems as well as ordinary discomfort for the employee. Asbestos should be used only under the strictest controls as OSHA has declared it a carcinogenic agent. If toxic gases controlled by OSHA standards, such as sulfur dioxide and carbon monoxide, are present in the facility, periodic measurements are required and records must be maintained. The facility inspection should note any facility deterioration that could cause leakage of any toxic gases. Walking and working surfaces should be checked to ensure that they are not slick from grease, oil, water, or other friction robbing materials.

Preventative Maintenance Check List for Fire Truck

Inspections	Interval	Dec. 1976	Mar. 1977	June 1977	Sept. 1977	Dec. 1977	Mar. 1978	June 1978
Lub chassis	3 mo							
Lube compt. latches	3 mo							
Change oil	3 mo							
Change fuel filter	3 mo							
Change air filter	3 mo							
Change oil filter	3 mo							
Check fuel crossovers	3 mo							
Drain and flush tank and pump	3 mo							
Drain water from fuel tank	3 mo							
Check and adjust slack adjustors	3 mo							
Check chassis for loose nuts and bolts	3 mo							
Check exhaust system	3 mo							
Check transfer case	3 mo							
Check rear end	3 mo							
Run house test on pump	3 mo							
Check fan belts	3 mo							
Check engine water pump	3 mo							
Check radiator and water hoses	3 mo							
Lube all levers and leakage	6 mo							
Change water filter	12 mo							

Initials of inspector _____

　　Remarks _____

Figure 7-4. *To ensure proper and reliable operation of a company fire truck, a comprehensive inspection should be conducted periodically.*

Fire Equipment Check List

Fire Pumps

Weekly Check

Date _____

Start and run pump for 10 minutes with pressure on system.

Quantity of fuel _____

Oil _____

Water _____

Battery Conditions _____

Leaks in system _____

Pump discharge pressure _____

Monthly Check

Date _____

Pump No. _____ Source of Power _____

Rated GPM _____ Pressure _____

Actual GPM _____ Pressure _____

Date of last fire pump GPM test _____

Does pump start automatically with pressure drop? _____

Fire Suppression Chemicals

Monthly Check

Date _____

Type _____

Amount _____

Condition _____

Do any chemicals need replacing? _____

Are extra chemicals needed? _____

Figure 7-5. *Fire pumps should be inspected on a weekly and a monthly basis, and fire suppression chemicals should be inspected monthly.*

Lunch rooms should be checked to ensure that chemicals from the work area have not ended up in the refrigerator. Microwave ovens should be checked for cleanliness. The microwave oven's electrical interlock should also be checked to ensure that radiation does not leak through the door or around the seals.

Electrical Inspection

The facility inspection should also include a detailed check of all electrical systems. All fuse and switch boxes should be secure, but easily accessible. All circuits should be marked as to what specific equipment or area they control. Circuits should be checked for proper grounding. A check should be made for unauthorized extension cords, or explosion-proof cords in areas where they are required. Check the integrity of the explosion proof system. If transformers are oil cooled, ensure they are serviced as prescribed by the manufacturer. Emergency lighting should be checked at each location. If the facility has an emergency generator it should be checked. Each facility should have a cutting and welding or "hot work" permit system for all work that is performed as maintenance outside the specific shop areas. Insurance company records indicate more fires are caused by cutting and welding operations than any other single cause. Therefore, it is of utmost importance that controls be maintained in areas of this work. If cutting and welding are observed during the facility inspection, the permit should be checked to ensure it was issued only after the specific checks were made.

The inspection should also include checks to ensure that all mechanical guards are in place and properly secured. Guards, such as electric eyes, dead man switches, and punch press guards, should be checked for proper operation. Two button switches that must both be depressed within a specific period of time should also be checked for proper operation. It is not uncommon to find some of these systems modified so that one button can be taped down, thus freeing one hand for speedier operation.

Showers and eye washes should be checked for proper operation. If portable eye washes or showers are utilized, make sure that this equipment is in its proper location and properly maintained.

No facility inspection can be considered complete until safety signs have been checked for adequacy of wording and location. Also, bulletin boards should be checked to ensure they are adequate and current.

The above areas of inspection depend on the specific facility and by no means should be considered complete. All OSHA and other governmental regulations should be consulted in order to develop a complete inspection list. These agencies also dictate the required frequency of inspection. This list is only a guide and should be used to stimulate ideas of what should be inspected in a particular facility.

8
Safety
Training

Safety training seems to be a catchall-cure-all phrase to a large portion of industry. There is a continual barrage of verbiage on the subject from the National Institute of Occupational Health (NIOSH) and the Occupational Safety and Health Administration (OSHA). Many standard and proposed standards require that employees be provided specialized training on a particular hazard and that records of this training be maintained. Most labor unions continually press for employees to be better trained on a particular job. It doesn't seem to matter whether the training is needed or not. Regardless of the need, it is good public relations for unions to press for additional training. When confronted by a problem in safety or other work areas, many employers start a training program before investigating if one is really needed.

All this training recommended or mandatory from three distinct forces, requires a monumental amount of man-hours and money. If the training is needed, and is properly prepared and presented, it is money well spent. However, if it is not needed, and is poorly prepared and presented, it is a needless economic drain on the employer.

Training and manpower development is probably one of the most important loss control programs that an employer can implement. Through correct training and development, injury accidents, noninjury accidents, fire, product damage, and all other asset grabbing losses can be greatly reduced. This, of course, requires that effective loss control procedures be integrated into the regular employee training.

Certainly safety training can and should be implemented without regular employee job training. But before any training is begun, it is important to determine if it is needed and what type should be used. Often, labor unions demand that specific types of training be provided. For example, some unions want employees to be trained as safety inspectors to perform duties above and beyond their jobs; some unions want employees trained to handle other or more sophisticated jobs solely because of seniority; and some unions want members to be trained as watchdogs, so that they can keep check on the company.

Loss Control Training

Training recommended by the employer may start at the highest level in the company. Many progressive companies engage training firms, which specialize in loss control, to conduct training sessions that last from one day to a week. Usually in this type of training the chief executive and all the top management employees including loss control management, will go to a retreat where job interference will not occur. Some of these sessions are very long and demanding, yet at the end of the session, whether it be one day or one week, the executives will probably be a lot better informed about safety and loss control measures.

If the loss control training is to continue in the company, the next level of management probably will be sent to a very similar course usually conducted by the same training firm. This type of involvement by management should be felt immediately throughout the work force.

These executives will probably come back with ideas and information that would require two or three years to completely implement. However, they will be taught to establish a course of action where progress of the safety program can be measured. The safety professional should play a leading role in developing a system that measures safety activity and performance for management use. This report should be prepared on a monthly basis and sent to all executives in the company. With this report management will become aware of any weakness in the safety program. If management has the proper attitude, correction of these weaknesses will be demanded. An astute safety professional will anticipate their desires and should be ready to make recommendations when the time comes.

The Importance of Supervisory Training

Another area of training that is vital to any organization is the supervisory training. Again, it is much preferable to integrate safety in the normal supervisory development plan. It is very desirable to train supervisors in safety, so that they may completely discharge their safety responsibility. The following are some areas that should be covered in special supervisory safety training courses:

1. How to train employees to work safely.
2. Accident investigation.
3. Job safety analysis.
4. How to conduct a safety meeting.
5. Facility inspections.
6. Accident causes.

The only one of the above subjects that will be discussed here is "How to train employees to work safely", as the rest of the subjects are covered in other chapters.

All supervisors face the task of having to train employees to work safely. To a few supervisors it is very easy, to most it is very difficult. Some companies go to specialized trainers for all their training regardless of the type. However, it is felt by many experts in the field that a competent supervisor can train his own employees much better than an outside trainer. However, on some specialized safety programs on equipment it may be desirable to bring in an expert in that field to do the training. When a company is deciding who should do their safety training, they should carefully look at their supervisors. If some of them are not qualified as trainers, it is a good idea to qualify them, if possible, before embarking on a training program. To develop supervisors into good trainers there are many important points to learn:

1. The instructor must become thoroughly familiar with the safety aspects, as well as with other pertinent knowledge, of the subject. When questions are asked about the subject, the instructor certainly should be able to have the correct answer.

2. Personality is probably as important as safety knowledge and general subject knowledge. However, many supervisors do not have a glowing personality. Yet, this can be overcome if the supervisor will put an honest effort in his instructing. He must show confidence and be able to control the training sessions. The instructor should also show enthusiasm for the course. If he does not, the course will probably be a dismal failure because the employees will feel that he doesn't believe in what he is doing. Tact is another trait that is very useful in employee training. Some people are naturally slow learners and one must be careful not to "turn them off" by being abrasive. Patience and understanding of the employees' problems with learning the subject matter is needed. If a supervisor is deficient in any of these qualities, he should strive for improvement, for they are the building blocks that make a great supervisor.

3. The supervisor in his instructing must watch the quality of his voice. He should pronounce his words as clearly as possible, and should use his voice to emphasize important points. Shop language or terminology should be used as much as possible in the training course. Call a piece of gear

by a slang name, give the proper name, then follow with the slang name and continue to use the slang throughout the presentation. Most likely the employees will retain more of what is taught if it is taught in the language that will be used in the shop. However, this statement does not mean that distasteful language should be used. The supervisor should by all means avoid monotones and lengthy reading. Certainly it is permissible to read short paragraphs, such as instructions on how to do something. Always try to maintain eye contact with students and avoid thought distracting mannerisms, such as continually scratching or cracking knuckles. This may not seem a real problem but it is amazing how college professors have some type of mannerism that makes it very difficult to keep thought processes intact.

4. It helps if the instructor can cite examples of what he is talking about that have occurred in the company's shops or in a nearby shop. Nothing gets attention on a safety point as quick as pointing out how someone was injured on a familiar piece of equipment.

5. Instructing will be much more successful if the instructor uses props. Charts, graphs, and cutaways used at the proper time is just what it takes to get a difficult point across. Movies, if applicable, are also useful tools. However, too many instructors are equipped with movies that are only remotely related to the subject.

6. Each training session should be divided up into lecture sessions and question and answer sessions. The instructor should know his employees well enough to know who will respond to questions. Those employees should be asked questions early in the question and answer session to stimulate the group.

Supervisors that plan to train employees should choose a classroom that has a comfortable temperature with good ventilation and lighting. The seating arrangement should be such that it is not crowded. Outside noise should be held to a minimum, and by all means, except emergencies, telephone calls or mini-conferences of someone dropping in for a quick discussion should not be allowed. The continuity of training can be quickly destroyed by such interruption.

Since so many OSHA training requirements require records to be kept, a roll should be taken and filed with a short summary of the subject matter discussed. An examination should also be given and kept as this is very good evidence to an OSHA inspector that employees have been adequately trained.

Since the Occupational Safety and Health Act of 1970 the Departments of Labor and Health, Education, and Welfare have continually increased their demands on industry to expand training. As standards are developed and issued, almost invariably training employees in the particular hazard recognition becomes mandatory. There are many types of training that OSHA has made mandatory, and controversial. Some of them are

1. If employees use respiratory protection, they must be fully trained; this includes knowing how to rescue a fellow employee from a toxic or oxygen deficient atmosphere. They must also know how to clean and maintain the equipment.

2. First-aid training has caused much controversy. Very few employers fail to recognize the benefits of a first-aid training program. The problem lies in the confusing way that OSHA defines how an employer determines if first-aid training is needed. Most safety professionals agree that first-aid training effectively reduces injuries because it implants a safety awareness in most employees. In fact, most concede that this reduction of injuries is, in itself, well worth the cost of the training.

3. If hearing protection is used, the employee must be trained in the use and fitting of hearing protection equipment. He also should be made aware of the need for such equipment.
4. Employees that drive motorized equipment such as fork lifts must be trained and certified as to their competency to operate the equipment.
5. Many standards require employees to be trained in how to recognize specific toxic hazards and how to protect themselves.

These are only a few of the growing number of safety standards with which an employer must comply. Certainly if the present trend continues, the list will be much longer. This is a problem which must be approached realistically. Obviously, everyone cannot become an expert in toxic hazards. As the list gets longer, the safety specialist must spend more time calming fears (many of which are unwarranted), answering unjustified union complaints, and trying to comply with various OSHA citations that were caused by petty complaints. Often, the complaint will not be justified, but, during the inspection, minor violations will be discovered, such as a handrail being slightly out of specification. The safety specialist's time would be better spent establishing safety programs or helping design equipment to eliminate the hazard than scurrying here and there trying to settle each and every complaint.

Regardless of who dictates that training be done, if it is attempted, it should be done properly. One must remember that safety training merely for the sake of fulfilling some regulations is unsound business practice. The safety professional should avoid such traps. If you plan to train, be sure you know why you are training and what you hope to accomplish. If you train, by all means keep a record of the training. An example of such a record is shown in Figure 8-1.

```
┌─────────────────────────────────────────────────────────────┐
│  CBA Company                      Dallas, Texas              │
│  Date  8-10-77                    Dept.        Freight       │
│  Type of training: Respiratory protection                   │
│  Equipment used: Self-contained air mask with pressure-     │
│                  demand regulator                           │
│  Length of training: 4 hours                                │
│  I certify that I, John Doe, received the above training    │
│  on the above date.                                         │
│                         Signature: _____  │
│  Instructor   Joe Roe   Signature: _____  │
└─────────────────────────────────────────────────────────────┘
```

Figure 8-1. *If any type of safety training is conducted, a record of it, such as this example, should be kept.*

9
Job Safety Analysis

A job safety analysis is a procedure to make jobs safe. The safety analysis identifies the hazards or potential accidents associated with each step of the job, determines the severity of the hazard, and tries to establish a probability of occurrence factor for an accident. The safety analysis also develops a solution that will either eliminate or control each hazard.

Selecting a Job for Analysis

When selecting a job to be analyzed one should first look into past accident frequency. Certainly, the jobs with the highest frequency should be the first to be analyzed. However, the severity of injuries should run a close second in determining which jobs are to be analyzed. In fact, many safety professionals use a frequency-severity combination. The safety professional should work with the supervisors responsible for the areas where the accidents occur. The frequency-severity determinations should be based on approximately 60% frequency and 40% severity. For example, suppose two specific jobs have ten accidents each in a given period. However, job A has four lost time injuries while job B has only one lost time injury.

Job A—.6 x 10 + .4 x 4 = 7.6
Job B—.6 x 10 + 1 x .4 = 6.4

As can be seen, job A should have a job safety analysis performed before job B. A supervisor who has both jobs under his supervision must have some rational way to make this determination, and this formula is a simple realistic approach to the problem.

The next category to be analyzed should be jobs that have high accident probability. This should be followed closely by jobs that have high severity potential. Again, the frequency-severity combination is a good way to look at these jobs. However, it must be realized that one does not have projected probabilities until the job safety analysis is completed. So if the frequency-severity combination is to be used, a certain amount of estimating is required.

Lastly, in selecting the job to be analyzed one should look closely at newly established or modified jobs. This is an excellent way to give the new job close observation. It not only will point out potential hazards but also will give management an excellent training tool.

Breaking the Job into Steps

The second stage of the analysis is to break the job down into steps. These steps should indicate what is done and the order in which it is done. It is very important to keep each step in sequential order. One step out of order can destroy the quality of the entire analysis by making the job appear very hazardous when, in fact, if the step is performed in its proper sequence, the job is completely safe. Figure 9-1 illustrates the necessary steps in conducting a job safety analysis.

Identifying Hazards

The safety analyst should determine if a person can strike against or make injurious contact with anything, such as hitting one's head on sharp objects. This part of the job safety analysis should also include injuries caused by protruding or falling ob-

Job Safety Analysis

Job: Removing Hydrocarbon Pump	Date: 8-1-77	Analysis by: J. Wilson
Job Title: Pump Repairman	Supervisor: E. Moore	Reviewed by: H. Brown
Department: Maintenance	Section: Pump Repair	Approved by: P. Williams
Required Personal Protective Equipment: gloves, safety hat, safety shoes, face shield		

Job Step Sequence	Potential Accident	Probability	Severity	Recommendation
1. Switch off power.	electrical shock	2	A	Do not stand in wet area. Ensure electrical is grounded.
2. Disconnect electrical components.	electrical shock	1	A	Lock out electrical, try to start pump.
3. Close valves on each side of pump.	muscle strain	2	C	If valves are tight, lubricate. Use correct body position.
4. Drain fluid from pump.	fire or explosion	2	A	Remove source of ignition. Be sure liquid is below auto-ignition temp. Have fire extinguisher available.
	thermal burn	1	C	Stand where liquid cannot splash on body.
5. Unbolt pump and remove.	skinned knuckle	1	D	Use proper tools.
	muscle strain	2	C	Use proper lifting equipment. Use correct body position.
	broken or bruised feet	2	B	Ensure lifting device has adequate capacity. Wear safety shoes.

Figure 9-1. A detailed job safety analysis, which associates each step in a job with its potential accident, is an effective tool for promoting accident prevention.

jects, determinations if the employee can be caught in, on, or between anything such as unguarded V-belts, pulleys, and gears, and investigations to see if an employee can fall in any way. Care should be taken to identify trip hazards, unguarded scaffolds, openings in walking surfaces, and slippery surfaces. Also, care should be taken to see if the employee can strain or over exert himself in any way. This would involve jobs that require more to be lifted than one person can safely handle, or jobs that require the employee to stand in an awkward or unsafe position while lifting.

Lastly, in identifying the potential hazards one should look for injurious environmental conditions to which the employee may be exposed, such as toxic fumes, lack of oxygen, particulate matter (dust and asbestos), heat, steam, and extreme cold conditions.

Accident Probability and Severity

Determining the probability of occurrence means establishing the likelihood of the hazard actually resulting in an accident. The probability factors are based upon the number of failures (these could be mechanical, human, or a combination of the two) required for an accident to occur and are broken down as follows:

0 —No failure required or failure occurred in advance. This indicates the highest likelihood of occurrence.
1.—One failure required.
2.—Two sequential failures required.
3.—Three sequential failures required.
4.—Four sequential failures required.

Normally this determination is only attempted in system safety analysis, yet if safety is to establish itself as a true management function, probabilities must be used. Too much hazard identification is presently based on the assumption that the severest possible accident will occur each time the hazard is encountered. This is not realistic because it does not take into

Job: Removing Hydrocarbon Pump	Analysis by: J. Wilson
Job Title: Pump Repairman	Date: 8-1-77
	Reviewed by: H. Brown
Department: Maintenance	Supervisor: E. Moore
	Section: Pump Repair
	Approved by: P. Williams

consideration the fact that a certain amount of unpredictable equipment failures will occur. These failures are a result of poor design or incorrect installation and are not related to the actions of the accident victim. Also, the failure may have happened prior to the sequence of events that finally caused the accident, and the victim either did not know the failure occurred or did not recognize the failure as a hazard. Obviously, an accident does not occur everytime someone of something fails. Safety rules can be violated, yet accidents do not occur. Enough failures do not occur in the proper sequence to actually trigger the accident. It is "near-misses" like these that undermine an employee's respect for a hazard, and result in tragic consequences.

It must be pointed out that many authorities feel that an accident cannot occur without a failure. Obviously, if a potential accident requires only one failure (whether it be human or mechanical), and the accident is given an A severity rating extreme preventive measures must be implemented. In actual practice the prudent company, upon discovering this potential would go to extreme measures to immediately eliminate such potentials. An accident that has a probability factor of four and an E severity rating should be dismissed from further consideration. Some will argue that it is not necessary to make determinations through four failures, but if one is to be assured of a reasonable amount of safety in probability determinations, the analysis must be carried through at least four failures.

Severity ratings are used to classify accident damage and are broken down as follows:

A—Fatality or severe financial loss.

B—Personal injury which likely would cause lost time, or large financial loss.

C—Personal injury requiring medical treatment, or medium sized financial loss.

D—Personal injury causing first aid only, or small financial loss.

E—Equipment damage only, with small financial loss.

It should be noted that each company must determine its own degree of financial loss.

Severity of accident potential determinations usually have not been made in the past. The safety profession has always, to some degree, had a stigma because too many safety professionals have recommended extreme protective measures where only a minor injury could occur. In many cases, even these would not occur if ordinary precautions were observed. It is easy for an individual to get caught in this kind of trap, if proper study and determinations have not been made of the potential hazard severity. If the safety profession is to obtain equal professional status with other company functions, a system of determining potential hazard severity should be used.

Areas of Accident Reduction

When potential accidents have been discovered, probabilities and severities determined, solutions to prevent the accident occurrence should be developed. The solutions will normally be found in one of the following areas: (1) job procedure, (2) job environment, (3) method change, (4) reduced frequency of exposure.

When changing the job procedure, one should look for steps or safeguards that can be inserted, which make the job a safe operation.

When trying to change the job environment one should look at the employees position in relation to equipment, roads, and physical hazards. Also, such hazards as noise, heat, extreme cold or toxic atmosphere should be controlled or eliminated.

When looking at the method change solution, one is looking at how a particular job is performed. The solution may be a revised way to feed a machine or grease a piece of equipment, or it might require a complete redesign of a piece of equipment.

When trying to change the job environment one should look at the employees' position in relation to equipment, roads, and considered to see if it is necessary where the exposure occurs. Does it need to occur as often as it presently occurs? Would protective equipment help in reducing the exposure? Sometimes the step where the exposure occurs is no longer needed as in the case where a worker was going onto a storage tank that contained H_2S gas once a shift. The worker was looking into the tank to determine the fluid level which was controlled by an automatic device. An investigation revealed that this procedure had been implemented many years earlier when the automatic level control was installed to make sure it was performing as desired. However, supervision simply forgot to eliminate the step after the system was proven reliable. Consequently, once each shift, three shifts a day, seven days a week for many years, employees were needlessly exposed to toxic quantities of H_2S gas. Fortunately, no one was severely injured. The point of all this is that jobs should be thoroughly checked periodically to see if steps or procedures need to be changed or eliminated.

Benefits of a Job Safety Analysis

After the job safety analysis has been completed it has at least four very useful functions other than eliminating job hazards. These are: (1) initial job safety training, (2) regular safety contacts, (3) pre-job safety contacts, and (4) cost improvement procedures.

A well-performed and maintained job safety analysis is excellent to use for initial job safety training. The analysis will have in written form each step of the job and the associated hazard. It also will tell the new employee how to avoid the hazard.

A job safety analysis that has been recently concluded, revised, or reemphasized makes an excellent subject for regular safety contacts of employees performing the analyzed job. It puts fresh or new emphasis on a job that may have contributed significantly to the accident picture.

Another valuable use of the job safety analysis is for pre-job contacts such as a five-minute safety meeting. It may be used to go through a job that is not performed regularly by the employees. It gives them a refresher course on each step and the associated hazards.

Lastly, the job safety analysis can be used for cost improvement studies. Studying the necessity of each step may lead to cost improvements through the elimination of unnecessary steps, equipment, or materials.

One job was analyzed in a plant that consisted of six different steps, each using a different chemical. As a result of the analysis two of the chemical steps were found unnecessary. The job was much safer by the two chemicals being eliminated and the company had a documented saving of $80,000 per year on this one job.

Methods of Performing a Job Safety Analysis

The job safety analysis can be performed by three different methods. They are: (1) observation method, (2) discussion method and (3) recall and check method.

When a supervisor observes a job being performed ideas are stimulated on how the job can be done safer or better. It also encourages the employee and the supervisor to discuss the safety of a particular job.

The discussion method is probably the preferred method. A group of supervisors and workers, and sometimes a safety department representative will sit down and discuss every aspect of the job. This method relies on the broad experience of everyone involved in the analysis. It also has the capability of silently training anyone in the group who may not know as much about the job as he should, and thus keeps that individual from asking seemingly embarrassing questions. The discussion method results will receive much wider acceptance, because the entire group has collectively analyzed the job and agreed on the hazards, probabilities, severity, and solutions.

The third method is the recall and check method. It is usually done by an individual supervisor who sits down and tries to recall each step of the job. After he has completely analyzed the job and made the necessary determinations, he should check with other supervisors to verify his findings. To obtain maximum results from this method it is much better for several supervisors to analyze the same job, then check their findings collectively. This method is least preferable of all three, because one must rely on memory instead of observing or discussing the job.

The job safety analysis can be a valuable tool in any operation and it is highly recommended. One must realize that it takes considerable effort to conduct and maintain a job safety analysis program. If your operation is not ready for a complete program of job safety analysis, pick the most critical jobs for analysis. Use as many supervisors as possible in the analysis, and after it is completed use it as a training tool for supervisors and workers. The idea is to train everyone to look at a particular job and at least do a mental analysis if a written one does not exist. In the mental analysis the worker should think about each step and associated hazard. This method is not as good as a formal analysis, but by proper training improved safety performance can be accomplished.

10
Medical
Programs

Any company regardless of size should have some type of medical program. The medical program may range from a simple pre-employment physical to a sophisticated program that includes annual physicals, consulting nurses, physical exercise facilities, weight reduction, or non-smoking incentives. The range of medical programs in different companies is almost limitless.

The safety performance of most companies is partly dependent upon the medical program. OSHA standards have caused much greater coordination between the medical and safety departments. Some companies have even placed their entire loss control department under the medical department. Regardless of how the company is organized there must be teamwork between the medical and safety departments.

The OSHA law requires a company to retain a physician, either on a full-time or part-time consultant basis. The doctor also must play an important role in industrial hygiene programs and have direct access to top management to ensure compliance with medical recommendation regarding job placement, transfers, working conditions that constitute a health hazard and return to work authorization after illness or injuries.

Many companies are simply too small and there are not enough doctors available for each to have a private physician. To overcome this some companies form mutual aid groups and employ one physician as a medical consultant. This type of program may not be ideal, but if a company cannot employ a full-time physician, then this arrangement can be effective. In addition to pre-employment physical and the treatment of occupational injury and illness, the part-time physician should also establish preventive health measures and provide proper consideration in job placement.

The safety department can be a great help in providing the physician with the physical requirements necessary to perform each particular job in a plant. The safety department, working with each operating department should determine these requirements. Too often good workers are turned down for employment simply because the examining physician does not know what physical requirements are required for a specific job. They have not been furnished these requirements and due to demands on their time do not have an opportunity to visit the facility to help make such determinations. Therefore, it is advantageous to have a form which lists the physical demands and working conditions for each job. Such a form is shown in Figure 10-1.

The pre-employment physical normally has five main purposes:

1. To place the prospective employee in a job that he is physically fit to perform.
2. To provide the employer information about what work the prospective employee is physically and mentally qualified to perform.
3. To ensure that the prospective employee does not introduce a communicable disease into the work environment.
4. To ensure that the work to which he will be assigned does not aggravate or complicate a preexistent condition.
5. To advise the employer of preexistent conditions so that excessive liability can be avoided.

Job Description

Plant _____   Dept._____

Job Title _____

Physical Demands

Check   Check

_____ 1. Crawling   _____ 13. Standing less
_____ 2. Crouching   than 50%
_____ 3. Climbing   _____ 14. Sitting 100%
_____ 4. Hearing   _____ 15. Seeing depth
_____ 5. Lifting 10-25 lbs _____ 16. Seeing color
_____ 6. Lifting 25-75 lbs _____ 17. Walking 100%
_____ 7. Lifting 75-125 lbs _____ 18. Eye protection
_____ 8. Pushing-pulling   needed
_____ 9. Reaching   _____ 19. Respiratory pro-
_____ 10. Raising right arm   tection needed
             above shoulder   _____ 20. Use of both hands
_____ 11. Raising left arm   _____ 21. Use of right hand
             above shoulder   only
_____ 12. Standing 100%   _____ 22. Use of left hand
                                               only

Work Environment

_____ 23. Hot   _____ 32. Head hazard
_____ 24. Cold   _____ 33. Inside
_____ 25. Humid   _____ 34. Outside
_____ 26. Dry   _____ 35. High noise 90
_____ 27. Toxic conditions _____   dBA or more
_____ 28. Dusty   _____ 36. Hand hazard
_____ 29. Oily   _____ 37. Burn hazard
_____ 30. Wet-water   (Thermal or cold)
_____ 31. Foot hazard

Special job requirements (such as extra long hours, travel, etc.)
_____

Job Description _____

_____

Figure 10-1. *Placing an employee in a job is facilitated when the physical requirements and working conditions that the person will experience are known.*

When checking physical requirement use $\sqrt{}$ + if requirement is heavy, use $\sqrt{}$ − if requirement is light, and $\sqrt{}$ if requirement is ordinary.

This completed form should be on file with the medical and safety departments and the foreman where the particular job is performed.

If prospective employee cannot meet the physical requirements as listed above, employee should be rejected for employment or placed in a job where physical requirements can be met. Plant manager's approval must be obtained before placing an employee in a job where physical requirements are only marginally met.

Figure 10-1. *Continued.*

Each company should establish physical rejection criteria. Normally the company physician should make this determination by assessing the work environment and physical requirements of the job. Obviously, it does not take the same physical requirements to be an accountant as it does to drive a truck, so the type of industry and job should play a significant role in the rejection criteria.

Functions of a Plant Nurse

Any plant that is large enough should employ a full-time nurse. This person, if properly chosen, can be a tremendous asset in the safety performance of a company. The nurse is even more important if the company does not have a full-time medical consultant on location. It is a fact that many injuries will be relatively minor if properly handled at the time of the injury. Also, being witness to a great number and a wide variety of injuries, the nurse is a very good spotter of potential problem areas. It is not uncommon for a nurse to call the safety department and ask them to investigate a particular area because some particular injury or series of injuries has cast suspicion upon a work area.

Another important role of the company nurse is to treat injuries, under the supervision of a physician. Thus, minor injuries can be treated and the employee returned to work. If the employee is sent outside the plant, considerable time may be lost. There are many complaints, probably much more than justified, that the doctor lets the employee take off unnecessary time or the employee malingers and does not return to work after being released from treatment. Of course, there is one sure way not to have this type of problem and that is to stop the injury from happening. Also, the nurse can establish a visitation program, to visit regularly any employee that is off work from an occupational injury or illness. This is a very effective method of getting employees back to work once they've recuperated.

Normally, the nurse will also perform such duties, if the plant is not too large, as filing worker's compensation claims, necessary recordkeeping for all medical surveillance programs, audiometric testing programs, and safety glass fitting. This person has many duties and certainly has a place in most loss control programs.

Fulfilling OSHA Requirements

As mentioned earlier, with the advent of OSHA the medical responsibilities of a company continue to mount as each new health standard is promulgated. Each health standard has certain requirements for medical surveillance if a prescribed percentage of the threshold limit of a toxic substance is exceeded. These standards also include mandatory requirements that medical records of employees be retained for twenty or more years. If a plant has a hazard that is covered by these standards, it is imperative that they get in compliance and stay in compliance. The only way this can be accomplished is through an effective medical program.

Some companies must use respirators to prevent exposure of their employees to toxic chemicals, air contaminants, or oxygen deficient atmospheres. Here the medical requirements apply to possible exposure and the use of the respirator.

An example of these medical requirements is the Asbestos Medical Examination Requirements as given in the June 27, 1974 edition of the Federal Register 1910.93a.

1. General: The employer shall provide and make available at his cost, medical examinations.
2. Preplacement: Within thirty calendar days following first employment in an occupation exposed to airborne concentrations of asbestos, a comprehensive medical examination shall be given. As a minimum, a chest roentgenogram, (posterior-anterior 14 x 17 inches) a history to elicit symptomatology of respiratory disease, and pulmonary function tests to include forced vital capacity (FVC) and forced expiratory volume at one second (FEV 1.0).
3. Annual Exams: At least annually employees exposed to airborne concentrations of asbestos fibers shall receive a comprehensive medical examination. As a minimum, a chest roentgenogram (posterior-anterior 14 x 17 inches), a history to elicit symptomatology of respiratory disease, and pulmonary function tests to include forced vital capacity (FVC) and forced expiratory volume at one second (FEV 1.0).
4. Termination of employment: Within thirty days before or after termination an employee exposed to airborne concentrations of asbestos fibers shall receive a comprehensive medical examination. As a minimum, a chest roentgenogram (posterior-anterior 14 x 17 inches), a history to elicit symptomatology of respiratory disease, and pulmonary function tests to include forced vital capacity (FVC) and forced expiratory volume at one second (FEV 1.0).
5. Medical Records: Employer shall maintain complete and accurate records of all such medical examinations. Records shall be retained by employer for at least twenty years.

As can be seen from the asbestos medical requirements the demands of OSHA on companies and the medical profession are enormous. In many areas of the country it is very difficult to engage a physician or even a group of physicians to satisfy the medical requirements of a company. Obviously, OSHA is trying to force companies to engineer out all of their hazardous atmospheres and well they should. However, in many cases technology is not far enough advanced to engineer out all the hazards. Also, even if the hazards have been eliminated in the day to day work environment, an accident may cause an over exposure which again would require medical surveillance. It must be admitted that there is no easy solution to the shortage of qualified physicians to handle the work load created by OSHA.

Screening Companies

One type of solution that is being tried is to have medical work, such as audiometric testing and physical screenings, performed by companies specializing in this work. They normally will have doctors and audiologists on a staff somewhere that reviews each test or audiogram. If something abnormal is spotted in a particular employee's test, the company is notified so that further examination may be conducted. This is a reasonably new approach and it remains to be seen if the quality of this type of service can be maintained to assure professional medical surveillance for the company's employees being serviced.

If a company is planning to use the service of a screening company, there are several points that should be examined:

1. Insist on names of at least three other companies that will give the screening company a recommendation. Check out these recommendations.

2. Ensure that screening equipment is calibrated and certified in accordance with any OSHA standard applying to the particular screening.

3. Ensure that the operator of the equipment has the necessary training, certification, or license required for the particular screening.

4. What type of professional medical staff is used to review screening results? A physician, or audiologist if the screening involves audiometric exams, should be on staff.

5. What assurance do you have that the staff is adequate to review all the screening results collected by the screening company?

6. Where and how are records maintained? This type of service normally would use computers for recordkeeping. Duplicate records should be provided the company within a short period of time after each screening. This is to ensure the company has continued access to these records.

7. Cost of local professional testing, if available, should be compared with the screening company prices. Be sure that the contract price includes price per screening, travel costs, waiting time, and number to be screened.

8. Ensure the screening being performed will meet OSHA requirements, if it is being performed to satisfy OSHA standards.

After a screening program has been installed the local company physician should spot check the screening results on a regular periodic basis to ensure that continued quality work is being performed. It must be remembered that medical screenings that do not satisfy the local company physician, probably will not satisfy OSHA standards, which would render them worthless.

The time has arrived for management and labor to work together in the interest of occupational safety and health. Some workers are still trying to blame management for injuries or illnesses that are not job related. Occasionally, management tries to avoid responsibility for injuries and illness that occur on the job by demanding proof that they are job related. An effective medical program is probably the only way to solve this costly problem for both management and labor. Workers must learn to trust company physicians and in turn company physicians must earn this trust. Obviously, the closer management and the medical department work together on these problems the better the medical program will be, which will improve workers' health and minimize the drain on company assets.

11
Off-the-Job Safety

Including off-the-job safety in a company safety program can be an effective part of the overall effort of a company to reduce accidents and accident costs. Operating cost and production schedules are affected almost as much as by employees being injured off the job as by employees being injured on the job.

Off-the-job safety can best be handled as part of the regular company in-plant safety program. It should be designed to create interest for every member of the employee's family. If the off-the-job safety program does reach each member of the family, each family member will have a tendency to be more aware of hazards and hazard correction around the home or the non-work environment.

Each year national promotions are held on fire safety, safe boating, water safety, etc. It is very easy for companies to join in this type of program by developing their own or purchasing handouts that can be placed in checks or boxes near employee exits. When in-plant awards programs are used, the award can very easily be turned into a family affair such as a safety picnic or barbecue. This is one activity that can stimulate interest in off-the-job safety for several weeks. Interest will be stimulated by the original notice of the planned event and should continue until a few days past the event. Another excellent off-the-job program is for the company to sponsor the National Safety Council Defensive Driving Course. This can stimulate interest in better driving habits and in some states an extra motivator for employees to take the course is that they receive a 10% reduction in certain kinds of auto insurance for up to three years.

Off-the-job safety should also be promoted occasionally as part of the regular on-the-job safety meetings. Special attention should be given by management to this part of the meeting by providing handouts that can be taken home. These handouts can be such things as a home hazard recognition check list in Figure 11-1. It is not intended to present a complete home safety check list, but one can be devised from the ideas presented.

Many companies are hesitant to get involved in off-the-job safety for fear of being accused of interfering with their employees' private lives. However, this fear is not justified unless some very unusual circumstances are occurring at the time. Certainly it would be improper to begin an off-the-job safety program while attempts are being made by employees to be represented by a labor union.

Aside from the humane aspects, one must recognize the economic justification of an off-the-job safety program. In analyzing the cost of off-the-job injuries, it should be recognized that their cost can easily be as large or larger than the cost of on-the-job injuries. For instance, if a company has salary continuance, then it is actually paying two salaries to get the job done, the injured employee's as well as his replacement. Medical plans are affected by experience. The accident cost of the plan will certainly increase if employees and their families suffer an abnormal amount of off-the-job injuries. If an employee does not suffer lost time as a result of his off-the-job injury, the injury may render him only partially productive, and he may have to take off a few hours periodically to receive medical attention. Also, cost may be incurred in training another employee to take over for the injured. Lower productivity from other employees may be incurred while they discuss the injury.

Sometimes off-the-job injury costs are very difficult to obtain. However, if a company has a medical insurance program,

Home Safety Check List

	Safe	Unsafe
Entrances		
Free of slipping and tripping hazards	_____	_____
Adequate lighting	_____	_____
No loose steps or boards	_____	_____
Hall and Stairs		
Well lighted	_____	_____
Free of slip or trip hazards	_____	_____
Are toys routinely left in walking areas	_____	_____
Kitchen		
Does everyone know what to do if a pan fire occurs on the stove?	_____	_____
Step stools for climbing	_____	_____
Cabinets kept orderly	_____	_____
Pan handles kept turned away from stove front	_____	_____
Curtains loose in stove area	_____	_____
Household chemicals stored away from food and children	_____	_____
Everyone aware of hazards associated with electrical appliances	_____	_____
Electrical wiring in good repair	_____	_____
Bathrooms		
Non-skid surfaces available in showers and tubs	_____	_____
Electrical appliances away from water outlets	_____	_____
Medicines out of children's reach	_____	_____
Combustibles away from space heater	_____	_____
Living Room—Den		
Fireplace screens in place	_____	_____
Liquor in bar secured from children	_____	_____
Sufficient ashtrays	_____	_____
Area free of trip hazards	_____	_____
Bedrooms		
Fire escape routes planned	_____	_____
Lamp near bed	_____	_____
Floor free of trip hazards	_____	_____
Furniture away from windows to prevent children from falling	_____	_____

Figure 11-1. A good way to promote off-the-job safety is to provide each employee with a home safety check list.

Home Safety Check List

Basement or Garage

Furnace and water heater checked for safe operation	⎯⎯	⎯⎯
Flammables not stored in area	⎯⎯	⎯⎯
Tools properly stored	⎯⎯	⎯⎯
Power tools properly guarded	⎯⎯	⎯⎯
Electrical fuse or breaker box safe	⎯⎯	⎯⎯

General

Firearms and ammunition under lock and key	⎯⎯	⎯⎯
Large plate glass marked	⎯⎯	⎯⎯
Adequate electrical outlets throughout home	⎯⎯	⎯⎯
Emergency telephone numbers posted near telephone	⎯⎯	⎯⎯

Figure 11-1. Continued.

records may be checked for past history. If no past history is available, it should not be too difficult to set up a method of developing and recording these costs.

A very good way to start an off-the-job safety program is to send the National Safety Council's *Family Safety* magazine to each employee's home. This can be done at a very minimal cost. The beginning of such a program can be announced in the company house organ or in special mailings to each employee's home. The announcement of such a program should be well planned. Management must be prepared to promote and carry it forward if it is to be effective. Before the initial announcement is made, at lease six months of the program should be developed and ready for use. Advance preparation should continue after the program has been initiated to prevent hastily planned or missed portions. An off-the-job safety program must be as well planned and receive as much management support as the in-plant safety program if it is to be effective.

It should be remembered that many employees leave their in-plant safety training at the plant. If these employees are to be safe off the job, then an off-the-job safety program is necessary.

12
Awards Programs

Many safety professionals feel that an awards program is necessary for a complete safety program. This is an area of safety that is surrounded with controversy.

Opponents of awards programs feel that it is part of the employee's job to work safely and no award is necessary or justified. Certainly it is part of every employee's job to perform his work in a safe manner. However, whether an award is necessary or justified is the controversial area. Others feel that an awards program will lead to hiding of accidents or cheating on accident records. One major concern is that if accidents are hidden and not reported the injury will not be treated and may become infected or further injured.

Awards programs that are based solely on such tactics as safety bingo or trading stamps for consecutive days without a lost time injury are doomed to failure. As employees become aware that nothing is actually being done to improve unsafe conditions or prevent unsafe acts, the program will usually fail within a short time. The time it takes to fail depends upon how long it takes the manager of the program to admit defeat.

A Case History

A few years ago two plants of the same new construction were placed into production. The work performed, plant layout, and products manufactured were essentially the same.

Plant No. 1 based its safety program on a safety bingo program. Cards were distributed to each employee and a bingo number was drawn each day that the plant worked without a lost time injury. Worthwhile prizes were purchased and placed on display. Bingo cards were wiped clean anytime a lost time injury occurred or when someone won the designated prize for the specific game.

Plant No. 2 started its safety program based on supervisory safety training, employee safety meetings aimed at hazard recognition, safe attitude development, and a program to engineer out or correct safety hazards. Plant management was involved in the safety program.

For the first six months, Plant No. 1 had a zero frequency rate (no lost time injuries per year) while Plant No. 2 started with a frequency rate of 7.0 and gradually dropped to 5.0 at the end of six months. Suddenly Plant No. 1 experienced a lost time injury. The bingo cards had to be wiped clean just as several employees were about to win a prize. The bingo game was started over and ran for about three weeks until another lost time injury occurred. The end was near for safety bingo in this plant. The bottom fell out, all sorts of hidden injuries as well as new injuries became lost time. As an end result, Plant No. 1 had a yearly frequency rate of 8.5.

Plant No. 2 kept its steady decline and at year's end it had a frequency rate of 3.5 lost time injuries per year. This steady decline continued through the next year, although the frequency did not drop as fast as it did the first year.

Plant No. 1 after the first year was placed on the same type of program as Plant No. 2 and after a few months that were required to recover from the bad trip of bingo, it also experienced a steady decline in injuries. Both plants continue to have a successful safety program with a frequency much lower than the industry average.

Realistic Goals

Proponents of awards programs think they are desirable: if the program is based on an achievable goal, it complements a well-rounded safety program and the award is something that will motivate the particular group.

An award is usually recognition for superior performance or a job well done. If this is the case, what difference does it make whether the award is for safety, quality, production, or outstanding sales. The end result is the same, someone is rewarded for his efforts.

An awards program should be developed to motivate employees to discover their safety problems and the solutions to these problems. Motivation techniques should aim for employee involvement, job enhancement, and personal or group recognition. This type of program rewards employees for safe acts instead of handing out punishment for unsafe acts.

Many modern safety thinkers feel that punishment for unsafe acts does so much damage to a safety program that it will take many months to repair, if it can ever be repaired. Normally peer pressure is all that is needed to make a well-planned awards program successful.

Awards programs should not constitute more than 5% of the entire safety program effort. Program objectives should be to help develop safe work habits and safe attitudes, direct attention to specific accident causes, supplement safety training, let employees participate in accident prevention, provide a channel of communication between workers and management, and show management concern for safety. This channel of communication should eliminate any excessive filtering of employee desires to management, or management's desires to employees. An awards program should not be expected to take the place of training, compensate for unsafe conditions, or unsafe work procedures.

Types of Awards Programs

There are many types of awards that can be used and many award goals that can be established. The following are several types of award programs that may be utilized:

1. Housekeeping awards—This type of award usually is conducted with all departments in the plant competing against each other. The period of time for each award period varies from plant to plant; however, the award period on housekeeping should never be shorter than one month nor longer than six months. The type of awards given normally depends on the length of the award period. The award may vary from publicity in the plant newspaper to a gift for each member or an awards dinner.

2. Goat awards or negative awards—Sometimes this negative type of award is used to recognize supervisors with the worst safety or housekeeping record. The supervisor is given some symbol such as a goat or pig to keep on his desk until someone else wins the award. This type of award should be either on a weekly or monthly basis. This is a very good way to stimulate supervisors' competitiveness in safety.

3. Accident frequency reduction—This is probably the most commonly used safety award. However, the majority of these award programs are based on a certain period of time without a lost time injury, instead of a meaningful award such as a reduction in medical cases. The lost time injury base instills cheating, hiding of accidents, and understating accident severity. This, in too many cases, has an effect of eventually undermining the entire safety program.
   It is much better to plan out a program where each unit is competing against its own previous record, not someone else's record. A 10% or more reduction in OSHA medical cases over a six-month period is considered by many as the ideal type of award program. Whatever award is decided on

at the beginning of the award period, should be meaningful and promptly awarded. At the time of the award, management has an excellent opportunity to get across more safety messages or other company communications. This type of an award may be a nice gift, tickets to a ball game, or a safety dinner or picnic. Sometimes it is a good idea to let the winners of the award determine, within guidelines, the next award.

4. Individual safety awards—This type of award usually is based on outstanding individual performance, such as 5-, 10-, or 15-year work record without a lost time injury. Here the employee is competing against his own record. The more accident-free years he maintains, the harder he will try to remain accident free. Meaningful awards should be given individually when the employee achieves the predetermined goal. The awards should get progressively better for each consecutive award. For instance after five years a man might be given a nice ball-point pen, after ten years an engraved wallet. The point is to continually have a desirable award that the man may obtain. The frequency of the awards should be no less than five years and no longer than ten years

One weakness in this type of awards program is that an employee may work nineteen years and eleven months accident free and be ready for a twenty-year award when someone else's mistake causes him to have a lost time injury. To prevent this type of injustice, much preplanning must be done before commencing the program. The preplanning should establish the type of award, the frequency of the award, and all of the rules. A three-man committee should be establishes to review each lost time accident to determine if the injury was caused by someone else's action. If someone else's immediate action was not involved, then the accident should be charged against the employee and his award period must start over.

One good way to start this type of awards program is after the first year and each year thereafter give the employee a hard hat or badge decal, indicating the number of years without a lost time injury. Most employees become very proud of this inexpensive award.

5. Best safety suggestion award—This type of award should be run on a combination of monthly and yearly award. Each unit involved in the suggestion award establishes a three-employee judging team. They evaluate the monthly suggestions and decide which will do the most to improve safety in the unit. Here the award should be come sort of recognition, such as a write-up in the company newspaper. The next step is for the plant or facility to establish a five-employee judging team to decide which unit has turned in the best suggestions. The award should be more significant, such as privileged parking for the month and additional publicity.

   The final step in this program is to establish three members of management to judge the twelve monthly winners and decide the winner for the year. This type of award should be substantial, such as a color television or an expensive trip to some resort for the employee and their spouse. The main weakness in this award program is that if it is not established and run fairly, much bad will may be developed.

6. Safe driving awards—This type of award should be based on basically two categories, miles driven and exposure in these miles, such as heavy city traffic or open highway miles.

   The award itself may be anything from simple recognition to a substantial award.

Any awards program, whatever the type, must be run fairly, have meaningful awards, and the awards must be awarded promptly if the program is to be successful. A word of caution, don't start an award program unless you intend to run it right and continue it for an indefinite period. In other words, an awards program should be established and maintained by company policy.

Appendix

This appendix describes major organizations, groups, societies, and private commercial organizations where readers may obtain additional information about the various subjects in this book. The listing certainly does not attempt to cover every source that provided information on the subjects discussed in this book.

Service Organizations

The American Society of Safety Engineers (850 Busse Highway, Park Ridge, Illinois 60068). This society is a major source of safety reference material in the U.S. and Canada. Article reproductions from *Professional Safety*, or former titles published since 1961 are available directly from Xerox University Microfilm, 300 N. Zeeb Rd., Ann Arbor, Michigan. Also a selected bibliography of reference materials in safety engineering and related fields may be obtained directly from society headquarters.

Labour Safety Council of Ontario (Dept. of Labour, 400 University Avenue, Toronto, Ontario, Canada). The council is very active in Canadian safety affairs and encourages the reexamination and modernization of safety rules to help people live a full and useful life.

The American Conference of Governmental Industrial Hygienists (P.O. Box 1937, Cincinnati, Ohio 45201). will provide publications which give threshold limit values of airborne contaminants, threshold limit values of chemical substances, and documentation of the threshold limit values for substances in workroom air for a nominal fee. These threshold limit values are necessary when developing an industrial hygiene program.

National Safety Council (425 N. Michigan Ave., Chicago, Illinois 60611). The National Safety Council is the largest organization in the world that devotes its entire effort to the prevention of injuries. The National Safety Council publishes the *National Safety News, The Industrial Supervisor* and a variety of newsletters. It also publishes the proceedings of each year's National Safety Congress which is held in Chicago. The proceedings are a record of the presentations made at the congress. They can be very useful in developing a safety program.

American National Standards Institute (1430 Broadway, New York, New York 10018). ANSI is one of the agencies that provided standards for OSHA adoption. Standards developed by ANSI as well as by many other organizations are considered minimum standards.

National Fire Protection Association (470 Atlantic Ave., Boston, Massachusetts 02110). Most NFPA standards also were adopted by OSHA. They develop standards on a regular basis which may become law under OSHA. NFPA has more than 25,-000 members. Many of the members serve on committees for standard development.

Training Organizations

The National Safety Council (425 N. Michigan Ave., Chicago, Illinois 60641). The National Safety Council provides regular training schools at its headquarters in Chicago on variety of safety subjects. It also provides home study courses. For information check directly with NSC. NSC also sponsors Key Man courses usually held by local safety councils. These courses are aimed at the first-line supervisor.

The International Safety Academy (P.O. Box 19600, Houston, Texas 77024. ISA regularly conducts training on a variety of safety subjects. The subjects are taught for any level of safety specialist. They are held in Macon as well as other selected places around the country.

The Defense Civil Preparedness Agency offers an industry/business emergency planning course at its staff college in Battlecreek, Michigan. These courses are held periodically. For information write to Defense Civil Preparedness Agency, Liaison Services Division, Washington, D.C. 20301

The American Society of Safety Engineers (850 Busse Highway, Park Ridge, Illinois 60068). The society each year holds a Professional Development Conference at locations other than its home office. For dates, locations, and cost contact society headquarters.

Total Loss Control Training Institute (P.O. Box 865, Station B, Willowdale, Ontario, Canada, M2k 2RI). The training institute offers courses at various locations on total loss control. For information contact the institute at the above address.

Publications

Professional Safety is the official publication of the American Society of Safety Engineers (850 Busse Highway, Park Ridge, Illionis 60068). The publication is published monthly and contains many timely articles that may be used in boosting a safety program. The cost of this publication is presently $10.00 per year.

Industrial Hygiene Digest (Industrial Health Foundation 5231 Center Ave., Pittsburgh, Pennsylvania 15232) is a monthly publication and costs $65.00 per year.

Occupational Hazards (P.O. Box 91368, Cleveland, Ohio 44101) is published monthly and has many timely articles on safety. The cost of this publication is presently $18.00 per year.

Protection is the official voice of the British Safety Professional (Alan Osborne & Associates, Circulation Dept. 113 Blackheath Park, London S.E. 3 OHA). The cost of this publication is $12.00 per year for eleven issues.

Bibliography

This book cannot present all views on ways to improve safety performance. Additional sources of information are listed so that one may contact the specific publication for a copy of the requested material. Usually copies of this material are available at a small cost.

Suggested Readings

1. "Accident Cause Analysis," U.S. Department of Labor, *Bulletin 270*, Washington, D.C., U.S. Government Printing Office.
2. *Accident Prevention Manual*, 7th edition, National Safety Council, (425 N. Michigan Ave., Chicago, Illinois 60641).
3. Allison, W.W., "Safety Motivation (By Example)," *National Safety News*, September 1974, pp. 70-73.
4. Berg, Arnold F., CPCU, "Total Safety from a Risk Manager's Viewpoint," *ASSE Journal*, January 1974, pp. 21-26.
5. Berry, Clyde M., PhD., "What Is an Industrial Hygienist?" *National Safety News*, August 1973, pp. 69-75.
6. *Best's Safety Directory*, 5th edition, A.M. Best.
7. Bird, Frank E. Jr., "Damage Control—A New Horizon in Accident Prevention," Coatsville, Pennsylvania Lukens Steel Company.
8. "Communications for the Safety Professional," National Safety Council, (425 N. Michigan Ave., Chicago, Illinois 60641).

9. Cordtz, Dan, "Safety on the Job Becomes a Major Job for Management," *Fortune,* November 1972, pp. 112-117 and 162-168.
10. "Corporate Safety, OSHA Lends Impetus," *Occupational Hazards,* April 1973, pp. 94-96.
11. "Eight Point Plan Pays Safety Dividends," *Occupational Hazards,* September 1973, pp. 42-45.
12. Fletcher, John A., "Total Loss Control," *ASSE Journal,* January 1974, pp. 16-20.
13. Ford, Margaret Custer, "Why Awards Work," *National Safety News,* September 1971, p. 48.
14. Gausch, John P., "Loss Control and the Organized Safety Effort," *National Safety News,* October 1972, pp. 84-88.
15. *Handbook of Occupational Safety and Health,* National Safety Council, Chicago, Illinois.
16. "How to Conduct an Accident Investigation," *Occupational Hazards, Executive Report,* Fall 1972, pp. 71-72.
17. "Incentives for Safety," *National Safety News,* September 1971, pp. 45-47.
18. "Inspections—Safety's Bulwark, How to Go About It," *Occupational Hazards, Executive Report,* Fall 1972, pp. 65-66.
19. Linane, James J., AB, MBA, "Managing the Loss Control Function," *ASSE Journal,* April 1973, pp. 17-21.
20. Lindsey, Walter W., "Who's a Head-Shrinker? Or Humanizing Supervisor-Employee Relations," *National Safety News,* February 1972, pp. 88-90.
21. Manuele, Fred A., CSP, P.E., "A Concept of a Total Loss Control System," *ASSE Journal,* June 1973, pp. 12-17.
22. Manuele, Fred A., CSP, P.E., "Successful Safety Programs," *Professional Safety,* December 1975, pp. 10-15.
23. Martin, W.A., "Inspections, Surveys and Audits," *National Safety News,* November 1971, pp. 50-53.
24. Martin, W.A., "Management and Accident Prevention," *National Safety News,* September 1975, pp. 81-82.
25. McPherson, Sam and Petersen, D.C., "Safety by Objectives," *National Safety News,* October 1973, pp. 66-69.

26. Micheal, George W., "Selling Your Accident Prevention Program to Management," *Security World*, March 1973, pp. 68-70.

27. Mims, Albert, "Are Safety Committees Useful," *Job Safety and Health*, March 1974, pp. 22-23.

28. Mims, Al, CSP, "Two Types of Safety Committees," *National Safety News*, January 1973, pp. 61-64.

29. "Motivating for Safety," *National Safety News*, February 1974, pp. 60-61.

30. "Motivation or Looking at Safety from the Inside Out," *National Safety News*, February 1971, pp. 40-41.

31. Naquin, Arthur J., B.E., CSP, "The Hidden Cost of Accidents," *Professional Safety*, December 1975, pp. 36-39.

32. Oresick, Andrew, CSP, "The Use, Care, and Inspection of Hand Portable Fire Extinguishers in Industry," *Professional Safety*, March 1976, pp. 32-39.

33. Pardee, John M., "The Role of the Safety Professional, A Drama of Infinite Acts," *National Safety News*, January 1976, pp. 46-51.

34. Peters, George A., CSP, "Systematic Safety," *National Safety News*, September 1975, pp. 83-90.

35. Petersen, Dan C., "Identifying Safety Training Needs," *ASSE Journal*, March 1973, pp. 19-24.

36. Petersen, Dan C., *Techniques of Safety Management*, McGraw-Hill.

37. Petersen, Dan, CSP, "The Future of Safety Management," *Professional Safety*, January 1976, pp. 32-39.

38. Pollock, Ted, *Managing Creatively*, Vols. I and II, Cahners Books, Boston, Massachusetts.

39. Pope, William C., "Safety and System Management," *National Safety News*, May 1971, pp. 56-57.

40. Rao, K.A., "Safety Motivation Begins at the Top," *National Safety News*, February 1975, pp. 57-59.

41. Redmond, William R., "Managing for Safety," *Professional Safety*, December 1974, pp. 39-42.

42. "Re-energize Your Safety Program with Group Dynamics," *Occupational Hazards,* September 1973, pp. 47-50.

43. Rosenfield, Harry A., "Safety, A People Problem," *National Safety News,* December 1973, pp. 54-56.

44. "Safety Contest Evaluation," *ASSE Journal,* May 1972, pp. 38-39.

45. *Safety Education,* 3rd edition, McGraw-Hill.

46. "Safety Education: What to Cover, How to Convey It," *Occupational Hazards, Executive Report,* Fall 1972, pp. 57-58.

47. "Safety Effectiveness Depends on Communication Effectiveness," *Construction Methods & Equipment,* February 1974, p. 27.

48. "Safety Meetings—Channeled for Action," *National Safety News,* January 1972, pp. 52-55.

49. "Science Sheds New Light on Accident Proneness," *Occupational Hazards,* September 1973, pp. 61-64.

50. *Selected Readings in Safety,* Academy Press.

51. Smith, L.C., CSP, "Fit Training Methods to Objectives," *National Safety News,* January 1972, pp. 56-57.

52. Smith, L.C., CSP, "Packaged Training Programs," *National Safety News,* January 1975, pp. 65-67.

53. Spence, Stanley F., "Job Safety Analysis and the Supervisor," *National Safety News,* February 1971, pp. 42-45.

54. "Take the Extra Step for Fire Protection," *National Safety News,* March 1976, pp. 106-114.

55. "Take the Extra Step for Maintenance and Inspections," *National Safety News,* March 1976, pp. 72-84.

56. "Take the Extra Step for Occupational Safety and Health," *National Safety News,* March 1976, pp. 69-70.

57. Tanke, T.J., "Organizing an Accident Prevention Program for a Construction Company," *National Safety News,* October 1975, pp. 87-89.

58. "The Foreman's Role: Pivotal to the Program's Success," *Occupational Hazards, Executive Report,* Fall 1972, pp. 61-62.

59. *The Guide to Occupational Safety Literature,* National Safety Council, Chicago, Illinois.

60. Underwood, H. Clark, "A Manager's Safety Commitment," *National Safety News*, September 1975, pp. 77-80.
61. Vervalin, Charles H., ed. *Fire Protection Manual for Processing Plants*, 2nd ed. Houston: Gulf Publishing Co., 1973.
62. "What to Look for in a Safety Manager," *Occupational Hazards, Executive Report*, Fall 1972, pp. 75-76.
63. Wolenz, George J., CSP, "Now is the Time for a Management Division," *Professional Safety*, April 1976, pp. 44-45.
64. "Your Pivotal Role in Loss Control Management," *Occupational Hazards*, August 1975, pp. 27-30.

Index